CONCEPTUAL DESIGN OF THE INTELLIGENT TRANSPORT SYSTEMS PROJECT—CASE IN GUI'AN NEW DISTRICT

DECEMBER 2019

 Creative Commons Attribution 3.0 IGO license (CC BY 3.0 IGO)

© 2019 Asian Development Bank
6 ADB Avenue, Mandaluyong City, 1550 Metro Manila, Philippines
Tel +63 2 8632 4444; Fax +63 2 8636 2444
www.adb.org

Some rights reserved. Published in 2019.

ISBN 978-92-9261-904-6 (print), 978-92-9261-905-3 (electronic)
Publication Stock No. TCS190561-2
DOI: http://dx.doi.org/10.22617/TCS190561-2

The views expressed in this publication are those of the authors and do not necessarily reflect the views and policies of the Asian Development Bank (ADB) or its Board of Governors or the governments they represent.

ADB does not guarantee the accuracy of the data included in this publication and accepts no responsibility for any consequence of their use. The mention of specific companies or products of manufacturers does not imply that they are endorsed or recommended by ADB in preference to others of a similar nature that are not mentioned.

By making any designation of or reference to a particular territory or geographic area, or by using the term "country" in this document, ADB does not intend to make any judgments as to the legal or other status of any territory or area.

This work is available under the Creative Commons Attribution 3.0 IGO license (CC BY 3.0 IGO) https://creativecommons.org/licenses/by/3.0/igo/. By using the content of this publication, you agree to be bound by the terms of this license. For attribution, translations, adaptations, and permissions, please read the provisions and terms of use at https://www.adb.org/terms-use#openaccess.

This CC license does not apply to non-ADB copyright materials in this publication. If the material is attributed to another source, please contact the copyright owner or publisher of that source for permission to reproduce it. ADB cannot be held liable for any claims that arise as a result of your use of the material.

Please contact pubsmarketing@adb.org if you have questions or comments with respect to content, or if you wish to obtain copyright permission for your intended use that does not fall within these terms, or for permission to use the ADB logo.

Corrigenda to ADB publications may be found at http://www.adb.org/publications/corrigenda.

Notes:
In this publication, "$" refers to United States dollars.

Cover design by Jasper Lauzon

Contents

Tables and Figures iv
Foreword vi
Acknowledgments vii
Executive Summary xi
Chapter 1: Introduction 1
Chapter 2: Systems Engineering for Intelligent Transport Systems Projects 5
    Systems Engineering 5
    The Vee Diagram, Procurement, and the Overall Project Life Cycle 6
    Successful Application of Conceptual Design/System Engineering Process:
    New York City 10
    Kansas City Scout 13
**Chapter 3: Intelligent Transport Systems Conceptual Design—Gui'an
             New District Case Study** 17
    Step 1:   Develop Intelligent Transport Systems Systems Objectives 18
    Step 2:   Identify Stakeholders and Establish the Stakeholder Working Group 21
    Step 3:   Develop Regional Intelligent Transport Systems Systems Architecture
                with Customized Service Packages 22
    Step 4:   Allocate Customized Service Packages to Intelligent
                Transport Systems Projects 30
    Step 5:   Stakeholder Validation of the Customized Service Packages 39
    Step 6:   Publish Intelligent Transport Systems Systems Conceptual Design 40
Chapter 4: Intelligent Transport Systems Project Planning and Recommendations 41
    Intelligent Transport Systems Systems Conceptual Design Project Schedule 41
    Estimated Intelligent Transport Systems Project Design-and-Build Schedule 42
    Recommended Resources and Specialty Skills for Detailed Design 44
    Project Analysis 46
    Recommendations 47
Chapter 5: Conclusions 48

References 50
Appendixes
    1.   List of Stakeholders and their Intelligent Transport Systems Elements 51
    2.   Customized Services Packages for Gui'an New District 55
    3.   Potential Benefits of Selected Gui'an Customized Service Packages 65
    4.   Project Communications Requirements Estimates 67
    5.   Open Standards and Protocols Recommendations 69

Tables and Figures

Tables

1	Objectives and Associated Metrics for the Intelligent Transport Systems in Gui'an New District	19
2	Gui'an Intelligent Transport Systems Working Group and their Scopes	21
3	Intelligent Transport Systems Service Areas and their Service Packages in ARC-IT version 8.1	24
4	Objectives to Customized Service Packages Traceability Matrix for Gui'an New District	32
5	Schedule for the Project Subsystems	43
6	Special Resources and Specialties Recommended for the Detailed Design of the Projects	44
A2	Customized Services Packages for Gui'an New District	55
A3	Potential Benefits of Selected Gui'an Customized Service Packages	65
A4	Communications Requirement for Selected Gui'an Customized Service Packages	68
A5	Application Layer Open Standards and Protocols Associated with Relevant Customized Service Packages	71

Figures

1A	Systems Engineering Vee Diagram for Intelligent Transport Systems Projects	6
1B	Systems Engineering Planning on Vee Diagram – Regional Planning	6
2	Relationships between Framework and Regional Architectures, Intelligent Transport Systems Standards, and Intelligent Transport Systems Projects	7
1C	Systems Engineering Planning on Vee Diagram – Analysis	8
1D	Systems Engineering Planning on Vee Diagram – Development	9
1E	Systems Engineering Planning on Vee Diagram – Implementation and Testing	9
1F	Systems Engineering Planning on Vee Diagram – After Deployment	10
3	Customized Service Package for the New York City Real-Time Passenger Information Program	11
4	Deployment Network in Kansas City	14
5	Overall Functions of the Integrated Modeling for Road Condition Prediction System	15

6	Deployment of the Integrated Modeling for Road Condition Prediction System	16
7	Steps for Developing the Intelligent Transport Systems Conceptual Design	17
8	Generic Intelligent Transport Systems Elements in the Framework Architecture	23
9	Information Flow Diagram between Two Elements	27
10	Information Flow Context Diagram for Cultural Tourism Investment Equipment Transit Vehicle On-Board	28
11	Customized Service Package for Basic Traffic Surveillance	30
12	Intelligent Transport Systems Data Warehouse	39
13	Gui'an Intelligent Transport Systems Conceptual Design	40
14	Gantt Chart for the Gui'an New District Conceptual Design Development	41

Foreword

Strategy 2030 sets the course for efforts by the Asian Development Bank (ADB) to respond to the changing needs of the region until 2030. ADB's vision is to achieve a prosperous, inclusive, resilient, and sustainable Asia and the Pacific, while sustaining its efforts to eradicate extreme poverty. ADB will add value to the region through finance, knowledge, and partnerships. Among the seven operational areas that ADB focuses on, "making cities more livable" echoes the core of the smart transportation project in Guizhou Province in the People's Republic of China (PRC).

Gui'an New District (Gui'an) was established as a new city in January 2014 with the objective of making it the economic development driver in of Guizhou Province and western parts of the PRC. Its key strategies are to (i) encourage innovation and high-tech industrial development, (ii) conserve the natural environment for a healthy and green city, and (iii) foster eco-tourism. Transport plays a crucial role in the development of Gui'an in line with these key strategies.

ADB is supporting the PRC in order to achieve the development potentials of this new city and more. The project also focuses on Gui'an Direct Administrative District (GDAD), which is within the city's planned development area. It aims to develop GDAD as a smart and livable city in order to attract residents, businesses, and tourists. To achieve a smart and livable city, sustainable transport systems that are coordinated, integrated, and optimized should be provided. The design and installation of intelligent transport systems (ITS) in GDAD is one of the outputs of ADB's assistance. ITS includes the data processing and data communications systems that support the operations and maintenance of surface transportation modes to improve the safety and efficiency of transporting people and goods. ITS can save time, money, and lives if they are properly planned and implemented. ADB is embarking for the first time in high-technology solutions for this type of project financing.

This report aims to provide a basic understanding of how to plan, design, and implement ITS projects. Focusing on the ITS conceptual design (CD) for Gui'an, the report (i) introduces the systems engineering approach used in the technology project design and deployment; (ii) describes the results of the ITS CD for the city; (iii) elaborates the process followed to develop the specific ITS CD; and (iv) provides guidance for development of ITS project conceptual designs in other regions.

We hope this report is able to share the lessons learned in developing the project to further support initiatives in Asia and the Pacific, as well as in other regions, to make cities more livable through high-level technology interventions.

Amy S.P. Leung
Director General
East Asia Department

Acknowledgments

This publication was based on the technical assistance project Guizhou Gui'an New District New Urbanization Smart Transport System Development. Susan Lim, senior transport specialist, Southeast Asia Department, led and managed the project. Gloria Gerilla-Teknomo, senior transport officer, East Asia Department (EARD), provided technical inputs and coordinated the development of this publication. Overall guidance and supervision was provided by Sujata Gupta, director, Sustainable Infrastructure Division, EARD.

Robert Jaffe prepared the Gui'an New District Intelligent Transport Systems Conceptual Design Final Report, which is the basis of this publication.

The authors would like to express sincere gratitude to the Republic of Korea's e-Asia and Knowledge Partnership Fund, which funded this study, and the Bureau of Economic Development of Gui'an New District for the support during project processing.

The authors also thank the peer reviewers, namely Arun Ramamurthy, senior infrastructure specialist (Digital Technology), EARD; Ki-joon Kim, principal transport specialist, Sustainable Development and Climate Change Department; Seok Yong Yoon, principal public management specialist (e-Governance), Sustainable Development and Climate Change Department; and Jiaqi Ma, academic director, Greater Cincinnati Advanced Transportation Collaborative, University of Cincinnati.

Definitions

Term	Definition
Customized service package (CSP)	A CSP is a part of a regional ITS architecture that implements a specific ITS service or group of related services. The CSP diagram illustrates how a set of stakeholder elements share information to implement the service(s).
Framework ITS architecture	A generic and technology-neutral arrangement of generic stakeholder ITS elements, and the information flows that the elements share to deliver surface transport services, which are intended to improve transport safety and/or efficiency.
Intelligent transport systems (ITS)	A system that uses data processing and/or data communications to improve the safety and/or efficiency of surface transport operation and/or maintenance, which includes all modes of surface transportation.
ITS conceptual design (CD)	Includes a regional ITS architecture that is customized for the specific needs of a specific region; also includes some analysis of stakeholder element costs, system benefits, system communication requirements, and standards or protocols that should be used for the identified information flows of the regional ITS architecture.
ITS elements	ITS elements have specific functional requirements and data dependency requirements on other ITS elements. Stakeholders deploy ITS by implementing their ITS elements.
ITS Feasibility Study Report (FSR)	The ITS FSR describes the ITS deployment at a high level so that a high-level design (e.g., selection of ITS and information technologies) and detailed design (detailed specifications for the selected technologies) can be developed, which will be traceable back to the FSR requirements.
Regional ITS architecture	An ITS architecture that has been tailored and customized for the specific surface transport needs of a specific region.

Term	Definition
Stakeholder	An institutional entity that has responsibility for operating, maintaining, or using one or more ITS services.
Stakeholder representative	A person who represents the interests of a stakeholder.
Systems engineering (SE)	A system development process that is intended to keep projects on budget and schedule, and meets all the needs that are originally intended. SE does this by detecting defects early when they are less expensive to correct.
Systems Engineering Management Plan (SEMP)	The Systems Engineering Management Plan is a document that defines the specific SE activities (e.g., development and review of traceability matrixes, testing of components against specifications, verifying the ITS against requirements, and validating the ITS against objectives) that will be implemented in a project to manage risks by early detection of defects.

Executive Summary

Gui'an New District is planned to be the most dynamic growth pole in Guizhou Province and southwest People's Republic of China (PRC). It will be designed and built into a safe, orderly, smooth, accessible, low-carbon, sustainable, and smart city. Among the ways to achieve this growth include the (i) deployment of intelligent transport systems (ITS), which incorporates advanced information processing technology, data communication technology, electronic sensing technology, control technology, Internet of Things, and cloud computing methods in a big data center, among other technologies; (ii) promotion of environment-friendly public transport with the use of battery electric buses to support mobility; and (iii) advancement of the pilot study of the intelligent connected vehicles system, an emerging industry which makes use of the information and communication technology revolution.

ITS projects are complex due to many interacting components and a lot of information collected and analyzed. Not only are ITS projects complex, they also have high uncertainties in terms of project costs, schedule, and systems requirements at the onset. This report provides a basic understanding of how to plan, design, and implement ITS projects for a region to minimize the ITS project risks and meet the needs of users. It introduces the systems engineering approach to project design and deployment and describes the process followed to develop the ITS conceptual design (ITS CD) specific for the Gui'an New District. This document also provides guidance for development of ITS CDs in other regions.

The development of the ITS CD follows a six-step process:

1. **ITS objectives development.** Identification of ITS objectives is an important step to be able to later identify ITS services suitable for any city or region, in this case, Gui'an New District. ITS objectives should be based on the government's ITS plan. These objectives are further refined during stakeholder consultation.

2. **Stakeholder identification and consultation.** One of the most important attributes of a successful ITS project is to gain the engagement of a group of senior individuals that can speak for each mode or support service in a region, to validate the project objectives and ensure that the deployment of an ITS, consistent with the ITS CD, will meet their needs.

3. **Regional ITS architecture including customized service packages (CSPs) development.** The regional ITS architecture is central to conceptual design development. Once the objectives are in place, the framework ITS architecture is used as a menu of possible generic ITS services, and the selected ITS services from that "menu" are then customized based on the actual stakeholders in Gui'an and their actual stakeholder elements. A service package addresses a specific service and collects together different functional requirements, physical objects (systems and devices), and information flows that provide the desired service. A CSP illustrates what information stakeholder elements share to implement an ITS service.

4. **Allocation of CSP to ITS projects.** There are about 76 CSPs developed to meet all the Gui'an New District ITS objectives. Each CSP was allocated to an actual ITS project to be designed and then built.

5. **Stakeholder validation.** After the initial draft of the complete set of CSPs are prepared, the members of the working group will review each of the CSPs that their ITS elements participate in.

6. **Publication of the ITS conceptual design.** The ITS CD contains a very large amount of information that will be used by developers for preparing detailed design specifications so it is better to publish it in the internet for easier access. The ITS CD for Gui'an is published at http://www.consystec.com/guian/web/index.htm.

The development of the ITS CD gives careful consideration to all stakeholders involved in project planning and design. The CSPs inside the ITS CD carefully plots and illustrates what information each stakeholder element needs to share to implement an ITS service. This is an important step in developing an ITS project because this is the basis of the detailed design of the ITS project. The ITS CD forms a road map of ITS development, avoiding piecemeal approach of ITS solutions. It will make sure that even when technology changes, various ITS will still be functional and interoperable. It empowers the city or municipal government that is developing ITS to ensure the systems are sustainable (i.e., institutionally and technically interoperable) for the long run.

Chapter 1

Introduction

Background

Economic growth in the western region of the People's Republic of China (PRC) has been slower than in the coastal provinces, and there are many parts of the region that still have relatively low urbanization and infrastructure development. Guizhou Province is in southwest PRC and has a population of 35.8 million, of which 46% are urban. In 2017, Guizhou had one of the lowest rates of gross domestic product (GDP) per capita in the PRC—$5750 or 63.6% of the PRC average ($9042)— and its rural poverty rate of 7.8% was more than double the national rate (3.1%). Increased urbanization; better infrastructure; and improved connectivity to markets, employment opportunities, and social services can reduce the high incidence of poverty in Guizhou and its surrounding regions.

Recognizing the importance of reducing the development gap between the western region and the coastal provinces as well as urban–rural income inequality, the Goverment of the PRC planned to develop new cities and transport infrastructure projects to improve connectivity and economic development in the region. It also emphasized the importance of ecological protection and environmentally sustainable growth. Guizhou is in the Yangtze River Economic Belt and an important province in the development of the surrounding southwest and far west regions. The 13th Five-Year Plan for Economic and Social Development of the People's Republic of China (2016–2020) continues to emphasize these priorities.[1]

The State Council of the PRC established Gui'an New District (Gui'an) as a new city in January 2014, with the objective of making it the economic development driver in Guizhou and neighboring provinces. The key strategies for Gui'an are to develop a high-level technology innovation hub to attract talent and business, encourage innovation and high-tech industrial development, conserve the natural environment to ensure a healthy green city, and foster ecotourism. This is in line with the PRC's new direction on urban development, which places increasing importance on innovation and coordination, eco-friendly policies, and the significance of people-centered development and urbanization.

Context

The Gui'an planned area comprises 1,795 square kilometers (km^2), with a population of 0.73 million (2016 estimate) that is expected to reach 2.3 million in 2030. The Gui'an Direct Administrative District (GDAD) covers 470 km^2, comprising over 20 villages or towns located between and originally under the jurisdiction of Guiyang City (Huaxi District and Qingzhen City) and Anshun City (Pingba County and Xixiu District). GDAD forms the core of Gui'an, which will be expanded toward both south and west to cover the whole of the planned area. The population of GDAD was 330,000 in 2017. It is expected to reach 1.4 million (including 1.1 million local residents and 300,000 college students and tourists) by 2030.

---

1 Government of the PRC. 2015. The 13th Five-Year Plan for Economic and Social Development of the People's Republic of China, 2016–2020. Beijing.

Currently, the road network in GDAD is underdeveloped and transport connectivity is poor. Most of the roads between towns and villages are in poor condition, of low technical standard, and have inadequate transport facilities. GDAD's urban core has 17 trunk roads that need to be upgraded to improve access and connection within and between its four economic zones. Intermodal connectivity is another challenge, with links from the high-speed train station to the central area still underdeveloped. Road construction will play an important role in integrating land development, forming the functionality of the core area, and strengthening the kind of accessibility and connectivity in GDAD that will facilitate business and commercial development.

Public transport in GDAD has much room for growth. Only nine urban bus routes have been in operation since 2014, mainly plying four of the trunk roads. Areas without this coverage are covered in part by rural–urban long-distance passenger bus lines. Peak transport capacity in GDAD's university area is insufficient: bus services are unreliable and passengers experience wait times of 30-45 minutes. Moreover, there are no permanent public transport facilities with exclusive land permits, making it difficult to organize efficient public transport operations with infrastructure like bus interchanges and terminals. Improving public transport coverage and services through the creation of a comprehensive network that meets passengers' needs, therefore, is a key priority.

Currently, of 108 buses already operating in GDAD, most are powered by liquefied natural gas (LNG) and some are battery electric buses (BEBs). The existing charging facilities fail to meet the overall charging requirements, resulting in low BEB availability rates. Together, the number of BEBs and associated infrastructure and facilities need to be increased to improve BEB accessibility, availability, and reliability. Finally, to improve the accessibility to public transport in general, commensurate public transport information and communication systems linked to operators and users should be established.

Issues

As Gui'an is planned to be the most dynamic growth pole in the southwest PRC, it is designed and built to be a safe, orderly, smooth, accessible, low-carbon, sustainable, and smart city. Among the ways to achieve this growth include the (i) deployment of intelligent transport systems, which will incorporate advanced information processing technology, data communication technology, electronic sensing technology, control technology, Internet of Things, and cloud computing methods in a big data center, among other technologies; (ii) promotion of environment-friendly public transport with the use of BEBS to support mobility; and (iii) advancement of a pilot study of an intelligent connected vehicles system, an emerging industry which makes use of the information and communication technology revolution.

It is natural to expect that stakeholders (i.e., modal owners, operators, or maintainers) for each mode (and related support entities) in Gui'an will have their own intelligent transport system (ITS) to provide the ITS services needed by that mode. For example, modes in Gui'an can be buses, rail, cars, taxis, shared mobility vehicles, bicycles, and pedestrians; and related support entities are roadway operators, transportation information providers, police, fire, emergency management system, parking operators, and electric vehicle charging operators. At the same time, there will be many opportunities to share information between the modes and related service entities to gain safety and efficiency advantages. These information-sharing opportunities where the information crosses institutional boundaries are both opportunities and challenges.

In the worst case scenario, each institutional entity designs and deploys ITS on its own, and then later decide how to integrate the systems. At this point, it is guaranteed that there will be a large amount of costly rework needed to enable sharing these many instances of information crossing institutional boundaries. A better approach which directly manages this risk is to plan the system requirements and interfaces up front in a technology-neutral ITS conceptual design (ITS CD), reviewed and approved by each of the local stakeholders, so that they reach regional agreement on what systems each stakeholder will deploy, and what the information exchanges will be between each of the stakeholder systems, and how that information will be encoded (ideally using open standards). Only then can each stakeholder begin to design, build, and deploy their systems, following the ITS CD as it pertains to their systems, thus to build the overall Gui'an ITS with the desired institutional and technical interoperability to meet each stakeholder's needs.

Opportunities

The safe, reliable, convenient, and comfortable movement of people and goods is important for the operation and growth of cities. This is the primary objective of any transportation investment. ITS can save time, money, and lives if they are properly planned and implemented. ITS includes the data processing and data communication systems that support the operations and maintenance of surface transportation modes to improve the safety and efficiency for transporting people and goods.

There are many integration risks inherent in designing and deploying transportation systems involving new and complex technologies, and involving interoperation between transportation systems deployed by different institutions. Thus, the planning, design, and implementation of ITS requires a systematic approach that follows a globally accepted ITS architecture framework to develop an ITS CD, which meets the specific transportation needs of the Gui'an stakeholders.

The ITS architecture framework provides a systematic basis for planning ITS implementation, facilitate their integration using standardized interfaces for information sharing when multiple systems are to be deployed, and in this way help to ensure near- and long-term interoperability of stakeholders' systems. The ITS CD is an early phase of the design process, which outlines needs-based function and form of the ITS, while assuring institutional and technical integration of the individual stakeholder systems. It includes the design of interactions, experiences, processes, and strategies. The ITS CD defines the functional and performance requirements for the ITS elements.

This report discusses the process of developing a multimodal, multistakeholder set of interoperating ITS, which starts with a regional ITS CD. This ITS CD is intended to manage the inherent integration risks of designing and deploying multiple interoperating ITS by different institutional entities, while still meeting all the needs originally intended by the stakeholders.

This report also aims to provide a basic understanding of how to plan, design, and implement ITS projects. Focusing on the ITS conceptual design (CD) for Gui'an, the report (i) introduces the systems engineering approach used in the technology project design and deployment; (ii) describes the results of the ITS CD for the city; (iii) elaborates the process followed to develop the specific ITS CD; and (iv) provides guidance for development of ITS project conceptual designs in other regions.

This report includes five chapters that are organized to describe the process for developing ITS CD. Chapter 2 briefly explains the guiding principles behind the systems engineering process. The Vee model briefly describes the progression of steps in developing complex ITS projects from planning to execution until retirement. Chapter 3 follows the steps in developing the ITS conceptual design for the Gui'an New District project. It explains the development of customized service packages and how to allocate them to ITS projects. Chapter 4 describes the resources involved and benefits gained in developing ITS projects, citing the Gui'an New District project as an example. Chapter 5 concludes the report with lessons learned during project design.

Chapter 2

Systems Engineering for Intelligent Transport Systems Projects

Systems Engineering

Systems engineering (SE) is an interdisciplinary process that focuses on managing the risks in the design, deployment, and maintenance of complex interacting elements over their life cycles. The individual outcome will be a combination of components that work in synergy to collectively perform a useful function. The SE process involves the top-down development of a system's functional and physical requirements from a basic set of mission objectives. The purpose is to organize information and knowledge to assist those who manage, direct, and control the planning, development, and operation of the systems necessary to accomplish the mission (Sage, 1992).

The overall objectives of SE are twofold:

 i. to keep projects being procured on budget and on schedule; and

 ii. to meet all the needs/objectives (i.e., "scope") originally intended for the project.

SE does this by detecting defects early, when defects are less expensive to correct. The more complex a project, the more the need is for SE, because there are more things that can go wrong to challenge the project being procured, and later during operation and maintenance of the project. Errors at the beginning of a project, if undetected/uncorrected, become more expensive to repair as the project proceeds.

The development of an ITS conceptual design (ITS CD) at the very beginning of a regional transportation technology project is intended to avoid very costly interoperability (or "institutional integration") defects later in the project procurement, operation, and maintenance, thus contributing to the two overall objectives of SE. ITS projects tend to be extremely challenging due to:

 i. many modern and complex technologies involved, and

 ii. many stakeholder entities each with their own ITS elements that may need to be interoperable (share information) across institutional boundaries.

Sharing information may create risks if there are multiple development processes involved in the implementation of the regional ITS elements. Misunderstandings on which information will be shared and how information will be encoded can contribute to the risk, thus the intended ITS services will only partly work or not work at all.

Following a systems engineering process (SEP) may help substantially reduce the number and severity of defects in complex ITS as contemplated for the Gui'an New District.

The Vee Diagram, Procurement, and the Overall Project Life Cycle

Figure 1 illustrates the Vee SE diagram for generic ITS projects. The Vee diagram shows the progression of steps to go from a regional plan in the upper left of the diagram to a deployed system in the upper right of the diagram. Each step will be explained in the following sections.

Figure 1A: Systems Engineering Vee Diagram for Intelligent Transport Systems Projects

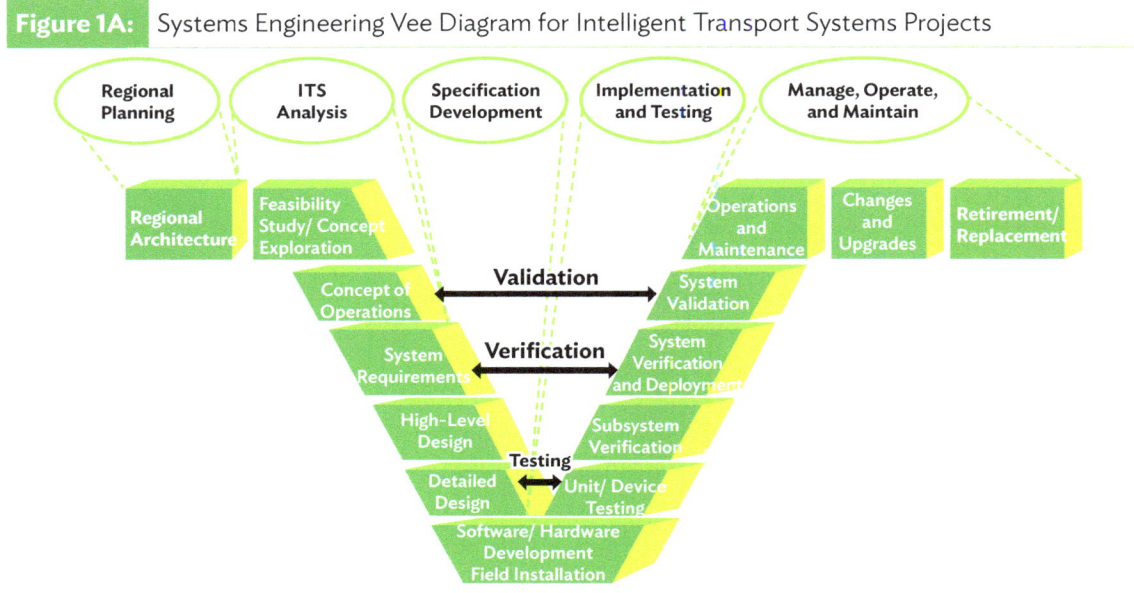

ITS= intelligent transport system.
Source: US Department of Transportation, et.al. 2007. *Systems Engineering for Intelligent Transportation Systems.* Washington, DC.

Regional planning. The first step in the SEP for ITS projects is to develop regional needs and specific objectives. Then, the regional ITS architecture is developed, which identifies the set of ITS services that will satisfy those needs and objectives. Also, the architectural solution to implement each ITS service is illustrated by identifying the stakeholder elements and their input and output information flows for each service in which they participate. The regional ITS architecture identifies the integration opportunities that should be implemented (and are agreed to by the stakeholders).

Figure 1B: Systems Engineering Planning on Vee Diagram – Regional Planning

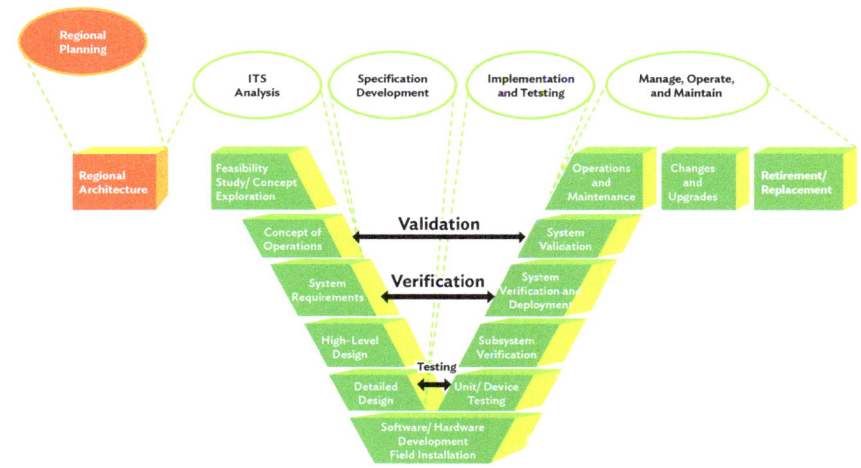

Finally, the regional ITS architecture includes identification of open standards (or local protocols) agreed to by the stakeholders, which will be used to encode the information flows (when those standards and/or protocols are available) between ITS elements. The relationships between the generic framework architecture, the specific regional ITS architecture, open standards, and the ITS projects for a region is illustrated in Figure 2.

The ITS CD includes the regional planning part of the Vee diagram as well as some of the ITS analysis steps in the next part of the Vee diagram. The ITS CD investigates how business needs and requirements and stakeholder needs and requirements are translated into a system-level understanding of the requirements. This understanding will inform what the system needs to do, how well it needs to perform, and what other systems it needs to interact with in order to meet the stakeholder and business needs and requirements.

Figure 2: Relationships between Framework and Regional Architectures, Intelligent Transport Systems Standards, and Intelligent Transport Systems Projects

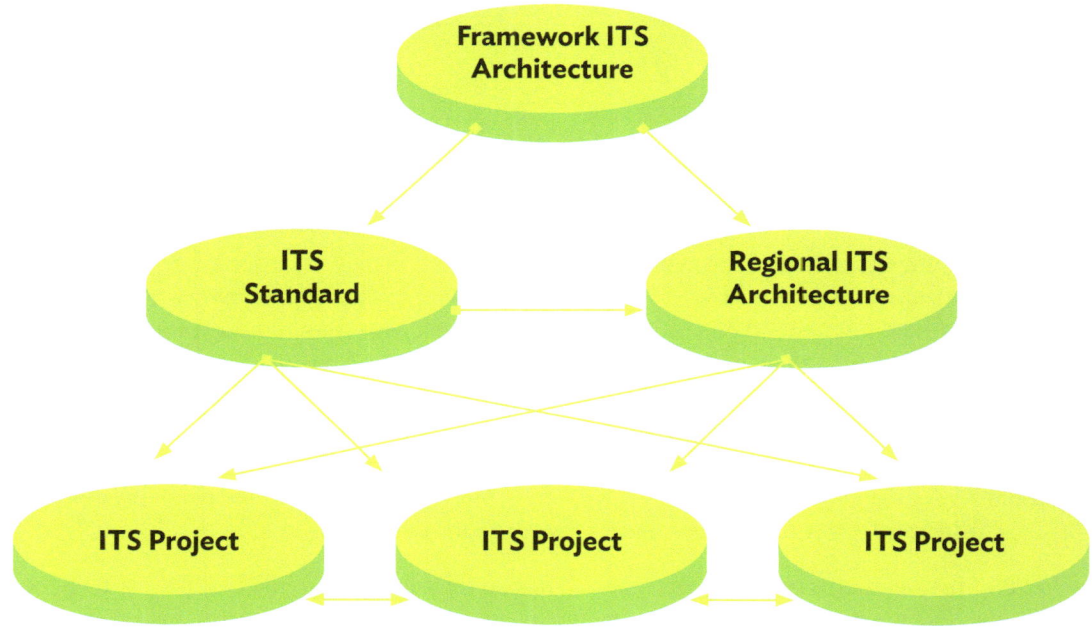

ITS= intelligent transport systems.
Source: Robert S. Jaffe.

ITS analysis. When complete, the regional ITS architecture should be used as input to the ITS Feasibility Study Report (ITS FSR) which should contain the ITS analysis. In the PRC, the ITS FSR contains the system requirements necessary to begin detailed design of the ITS projects in a region by the selected design institutes. The ITS FSR should include or reference the ITS stakeholder elements and their functional requirements as well as information dependencies on other ITS stakeholder elements in the regional ITS architecture. The ITS FSR should then add additional analysis such as detailed cost and benefits estimates, and high-level technology choices for the ITS elements guided by local environmental and institutional considerations. High-level versions of these analyses are included in the ITS CD. There should be full bidirectional traceability from the functional and information dependency requirements in the regional ITS architecture of the ITS CD to the ITS FSR. The FSR should also state the measurable objectives of the ITS, which can be used later to validate that the deployed ITS has satisfied all the objectives. (In the United States, many of the elements of an FSR are commonly part of two documents: the System Concept of Operations and the Regional Strategic Plan.) The ITS FSR is then used as input to the detailed design process, starting with a complete system requirements analysis.

Figure 1C: Systems Engineering Planning on Vee Diagram – Analysis

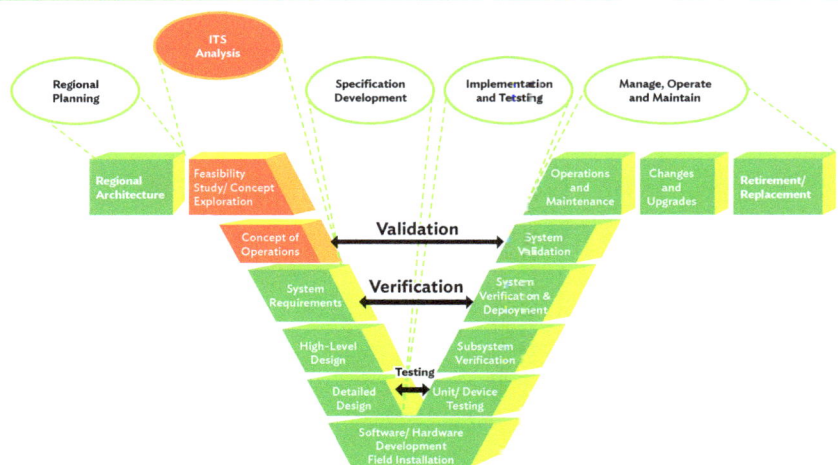

Requirements development. The system requirements state the functional, performance, and environmental requirements on each stakeholder project that makes up the overall ITS. The system requirements should not depend on the selected technologies. The system requirements should trace back to the high-level objectives in the ITS FSR and should be useful as verification tests on the final ITS before it is released to operations. The purpose of the system requirements (where every requirement is testable) is to verify that the system has been built correctly. This is complementary (but different) from the validation testing based on the ITS FSR, which should be used to confirm that the system meets the needs/objectives originally intended, i.e., the system requirements verify that the overall system is built right, and the ITS FSR needs/objectives validate that the right system was built.

The focus of the functional requirements in the FSR is on the functional impact to the users, operators, and maintainers outside the ITS, and the functional requirements of the system requirements focus on the functional operations of the ITS. For each objective in the ITS FSR, there should be traceability to one or more requirements in the FSR that, on inspection, persuade the reviewer that the ITS FSR objectives will be satisfied by the ITS requirements (and that, conversely, each system requirement contributes to satisfying at least one objective).

High-level design and detailed design. Once the ITS requirements (functional, performance, and environmental) are defined, they can be used to select the technologies to be used in the ITS (each selection supported by trade studies as necessary to justify the selection of the "best" technologies). This set of selected technologies are included in the "high-level design," not the system requirements. Next, the ITS detailed design proceeds, which includes the development of all the specifications necessary to implement the selected technologies for the full ITS. The detailed specifications are each tested in the integration of the ITS modules as they are built or procured. ITS modules can be either hardware and/or software.

Figure 1D : Systems Engineering Planning on Vee Diagram – Development

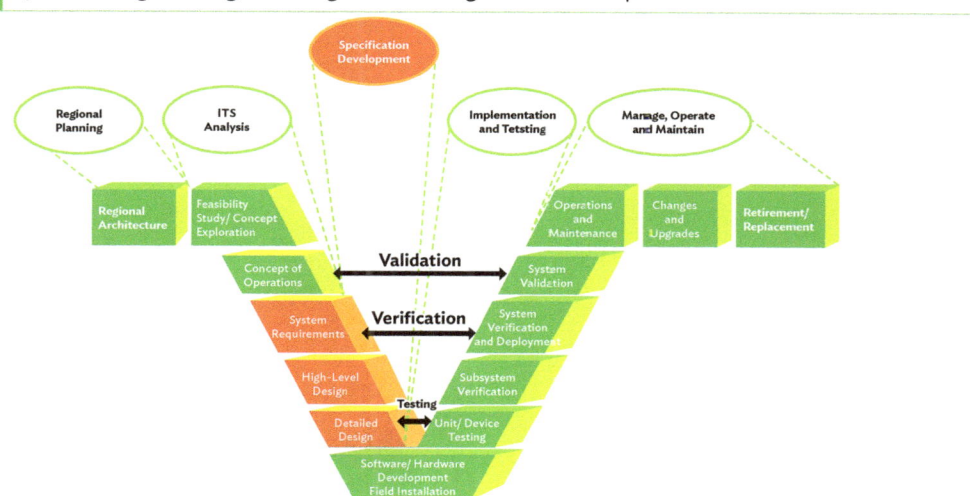

Implementation and testing. The left side of the Vee diagram is sometimes called the "decomposition" side, because it starts with a very high-level view of the ITS in the regional ITS architecture, and then gets more and more detailed as each high-level objective is decomposed into system requirements, and then specific technology specifications resulting in a complete detailed design for the ITS. The base of the diagram is the actual build or procurement of hardware and/or software for the ITS, and then the right side of the diagram begins, sometimes called the "recomposition" side, where the individual hardware elements and software modules of the ITS are integrated, tested and combined to result in the complete ITS, verified that the system was built correctly and tested to meet all requirements, and then put into operation and finally validated that the needs are all met.

Figure 1E : Systems Engineering Planning on Vee Diagram – Implementation and Testing

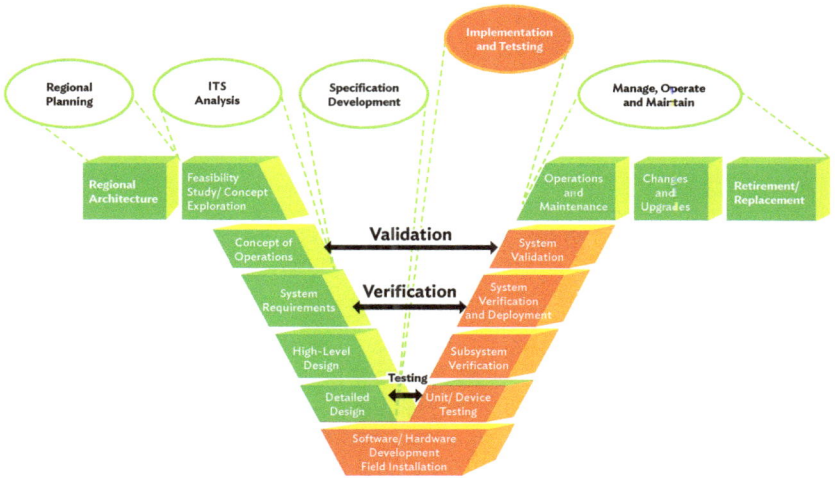

Manage, operate, and maintain. After deployment, the system can be "validated" that, in operation, it meets all the needs or objectives initially intended. After a period of operation, upgrades and changes may occur, and eventually replacement and/or retirement.

In proceeding from each stage to the next, the SE Vee diagram in Figure 1 shows a space between subsequent steps in the process, which represents a traceability analysis step that verifies that the subsequent level of analysis, design, or build, completely meets all the objectives, requirements, or specifications of the previous step. Further, there are three additional testing phases for unit/device testing against specifications, system verification against system requirements, and system validation against needs/objectives. All this traceability analysis and testing are intended to detect defects soon after they are created, when they are least expensive to repair.

Figure 1F: Systems Engineering Planning on Vee Diagram – After Deployment

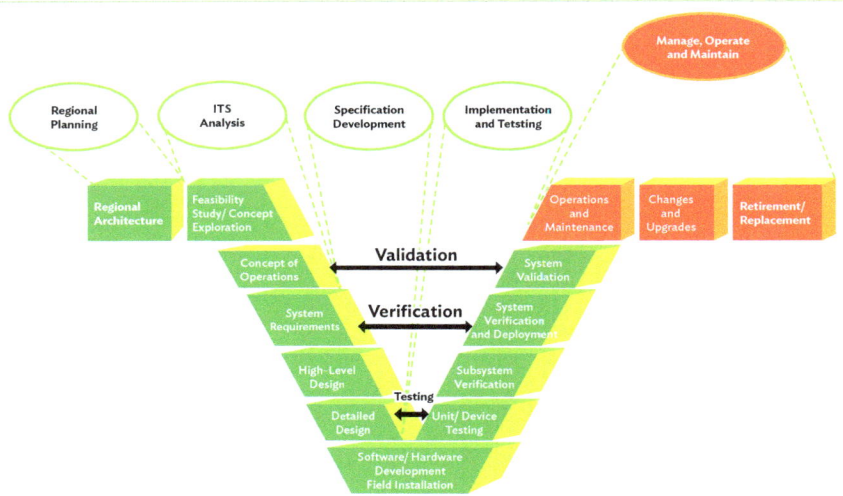

Finally, it is necessary to assess in advance if all the stages of the SE process are necessary for a project, based on where the risks in the ITS to be deployed exist. If there are a few risks in a stage of analysis (for example, because that stage of analysis is similar to a prior project that was successfully procured and deployed), then the traceability and/or testing of that stage might be unnecessary, or at least the analysis/testing can be abridged. These decisions should be made by staff experienced in SE and documented in a Systems Engineering Management Plan (SEMP) for the project. The SEMP should become a part of the Project Management Plan (PMP) or a parallel document to the PMP.

Successful Application of Conceptual Design/Systems Engineering Process: New York City

Under the leadership of New York State Department of Transportation (NYSDOT) Region 11 (which geographically corresponds to New York City), the transportation stakeholders of the region have developed an ITS conceptual design, specifically a New York City regional ITS architecture covering all modes of surface transport. For historical reasons, the city's regional ITS architecture is called the New York City subregional ITS architecture (NYCSRA). The four key stakeholders for this region are NYSDOT Region 11, which operates many of the regions limited access highways; New York City Department of Transportation (NYCDOT), which operates the surface streets and all ITS equipment on or near the non-limited access roadways and the East River bridges; MTA Bridges and Tunnels which operates major toll bridges and tunnels in New York City, and MTA Transit, which operates the subway and bus systems in New York City. Also included are many other transportation and public safety agencies.[2]

---

2 The regional ITS architecture is documented on a website that can be found at http://www.consystec.com/nycsra2018/web/index.htm.

All ITS projects in New York City are based on the conceptual design of the project in the NYCSRA. For example, a representative ITS project is the Real-Time Passenger Information (RTPI) Program. In this project, NYCDOT is responsible for replacing standard bus stop poles with RTPI displays both in a visual format and with push-button audible technology to provide information on bus routes and real-time bus arrival information. The potential program size for this RTPI work is between 8,000 and 14,000 standard bus stops. By the end of 2018, NYCDOT had installed RTPI signs at 339 citywide bus stop locations.

Figure 3 shows the customized service package (CSP) for the RTPI service. A CSP illustrates what information regional stakeholder elements share to implement an ITS service. The basic RTPI service, in this case, accesses real-time transit schedule information (architecture flow "transit service information") from the MTA-INFO Website (located at the Transit Management Center operated by MTA NYC Bus). This information is delivered on a pole-mounted digital display that shows when the next bus will arrive for each bus line serving the bus stop adjacent to the RTPI pole. The next bus information is also provided audibly when a vision-impaired traveler activates a button for this purpose on the pole. Other CSPs, not shown in this report, collect bus real-time automated vehicle location data in the Transit Management Center, and that set of data sources is used to estimate the real-time arrival information for buses for each stop along their route. Also shown in Figure 3 is the connection from the RTPI poles to an "NYC LinkNYC" kiosk. These are large digital message signs with keyboards that pedestrians can operate that are not necessarily located next to a bus stop. These large displays will collect information about bus arrivals in the neighborhood (as well as subway trains), so that pedestrians can consider their best options for transportation on bus or subway services in the area of the NYC LinkNYC kiosk.

Figure 3: Customized Service Package for the New York City Real-Time Passenger Information Program

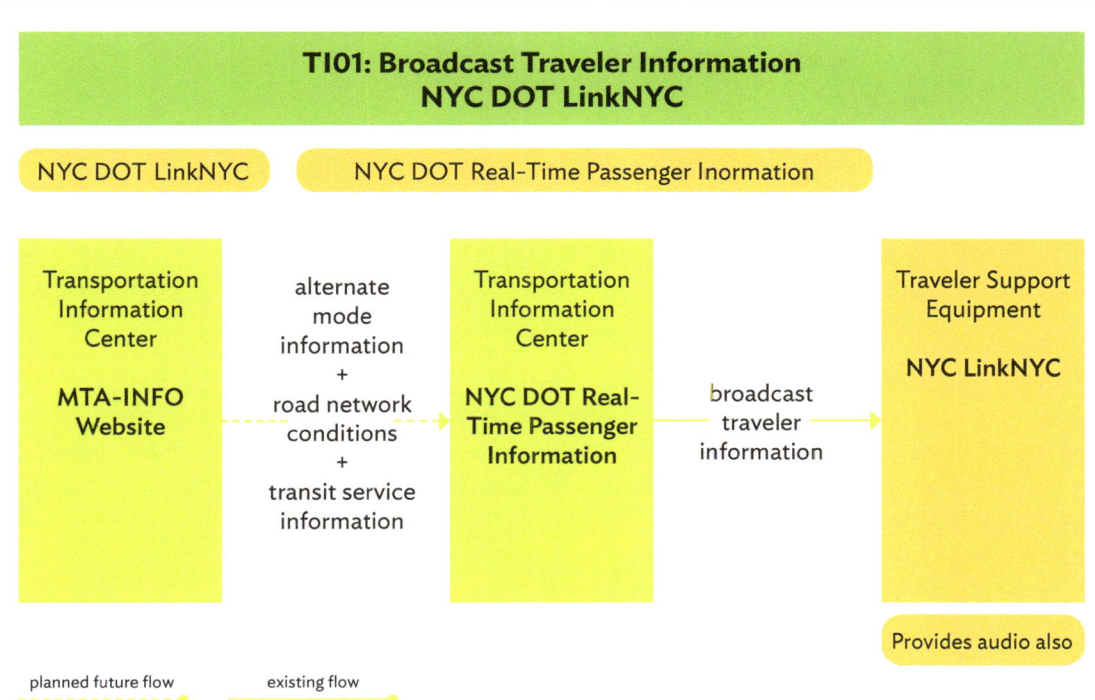

NYCDOT = New York City Department of Transportation.
Source: NYC Subregional ITS Architecture. http://www.consystec.com/nycsra2018/web/spinstance.htm?id=T101-02.

An actual NYC RTPI pole is shown in the photo. This particular bus stop has three bus routes stopping at it. The digital display shows the estimated number of minutes before the specific next bus arrives, with no display for the bus (M101) that is passing the stop at the moment the photograph was taken (and note that the bus did not pull over to the curb because there were no passengers waiting and apparently no passengers onboard requesting a stop).

At a bus stop. Real-time passenger information or RTPI is setup on a pole at a bus stop in New York City (photo by Robert S. Jaffe).

The normal project development process in New York City for a project with an ITS component is to first identify the part of the regional ITS architecture that represents the project and develop a project systems engineering analysis (PSEA). The PSEA, along with other technical and institutional requirements, is used as the basis for a detailed design of the project. Then the detailed design is used to "go out to bid" for construction, installation, and testing.

Successful Application of Conceptual Design/Systems Engineering Process: Kansas City Scout

While the New York City example illustrates well the benefits of applying system engineering process (SEP) at a region/city scale, another example in Kansas City shows how SEP can guide the development of a traffic management subsystem and benefit the entire traffic operations, demand management, and roadway maintenance in daily traffic management center activities[3].

Weather has a huge impact on safety, mobility, and productivity: annually in the United States, there are more than 1.5 million crashes, 600,000 injuries, and 7,000 fatalities occurring under adverse weather conditions. Delays from snow, ice, and fog cost about $11.6 billion per year. State transportation departments spend $2 billion per year on snow and ice control, and $5 billion per year on infrastructure repairs due to snow and ice. Road weather information is effective at alleviating the impacts of adverse weather: research shows that more timely, accurate, and route-segment or spot-specific weather and road condition information (e.g., alerts that are specific to where a traveler is or is going to be, rather than generic "watch for fog" information) can reduce weather-related crashes by changing driver behavior. Research also shows that decision support tools using road weather information will enable active management of surface transportation operations, promote effective and efficient roadway maintenance activities, and support better travel choices.

The Road Weather Management Program of the United States Department of Transportation (USDOT) has been working to promote safety, mobility, and productivity on the nation's surface transportation system by advancing road weather research utilizing observations from environmental sensor stations from road weather information systems; information from connected vehicles (CANBus and external weather sensors) from the integrated mobile observations projects in Michigan, Minnesota, and Nevada; creating the vehicle data translator to ingest, quality check, perform pavement and/or surface weather condition inferences, and disseminate the results; and establishing a research system, the weather data environment or WxDE, for the collection, quality checking, and dissemination of weather-related observations. In addition, the Road Weather Management Program has created two new weather-related application prototypes to convey information from the visual display terminal: the motorists advisories and warnings and the enhanced maintenance decision support system.

The Integrated Modeling for Road Condition Prediction (IMRCP) project was initiated by USDOT in 2015 and aims to create a tool that incorporates real-time and/or archived data and results from an ensemble of applicable deterministic and probabilistic forecast models (e.g., road weather, traffic, work zones, incidents) and fuses them in order to predict the current and future overall road/travel conditions for travelers, transportation operators, and maintenance providers.

Kansas City was selected as the deployment location, and therefore traffic management centers at Kansas City (KCScout) have been engaged as key stakeholders, along with other stakeholders across the US, to understand system concept design, requirements, and detailed design. A portion of the Kansas City metro area along a congested interstate corridor and surrounding arterials has been used for a demonstration study and evaluation area. The Kansas City area is subject to highly variable weather conditions and local recurring congestion typical of US urban/suburban settings. The I-435 corridor

---

[3] J. Ma et.al. 2017. *Integrated Modeling for Road Condition Prediction*. Final Report. https://collaboration.fhwa.dot.gov/dot/fhwa/RWMX/Documents/IMRCP/IMRCP_Final_Report%2012-20-17-FOR%20508.pdf.

along the southern part of the metro carries heavy commuter traffic in both directions and for much of its length runs along a stream way with historically significant flood risk. The corridor is well-instrumented for traffic, weather, and hydrology.

In the initial phase of the project, the Concept of Operations and Detailed Requirements were created. In phase 2, the System Architecture, System Design, Test Plans, Installation Guide, User Guide, and Marketing materials were created. Additionally, the IMRCP was deployed over highways and arterials in the southern part of the Kansas City Scout area. The study area is made up of 2,006 links, 870 nodes, and 188 bridges. Data in the study area are collected from 205 traffic signals, 105 traffic detectors, 53 ramp detectors, 15 dynamic messaging signs, 20 StormWatch sites, 5 Advanced Hydrological Prediction System stations, and an Automated Surface Observing System station. The deployment network in Kansas City is shown in Figure 4.

Figure 4: Deployment Network in Kansas City

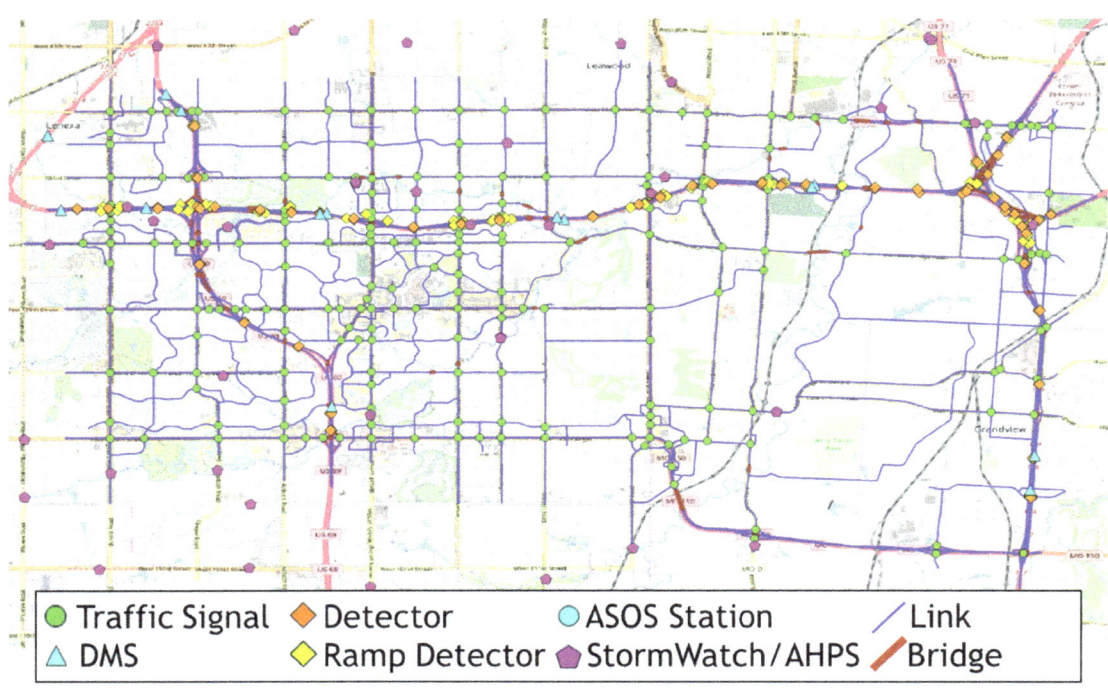

AHPS = Advanced Hydrological Prediction System, ASOS = Automated Surface Observing System, DMS = dynamic messaging signs.
Source: J.K Garrett et al. 2017. Integrated Modeling for Road Condition Prediction (No. FHWA-JPO-18-631). United States. Dept. of Transportation. ITS Joint Program Office.

Figure 5 depicts the overall functions of the IMRCP system. Inputs to the IMRCP system are listed on the left side of the figure. Data are currently being collected from sources that are color-coded in green– road weather data services, weather data services, Traffic Management Center (TMC) and advanced transportation management systems (ATMS), and traffic data services. All IMRCP system packages are shown inside the dashed box. Packages shown in green have been completed. The inner box contains forecast packages within the IMRCP. Users of the IMRCP are listed on the right side of the figure. Transportation operators have had access to the IMRCP as part of the evaluation process.

Figure 5: Overall Functions of the Integrated Modeling for Road Condition Prediction System

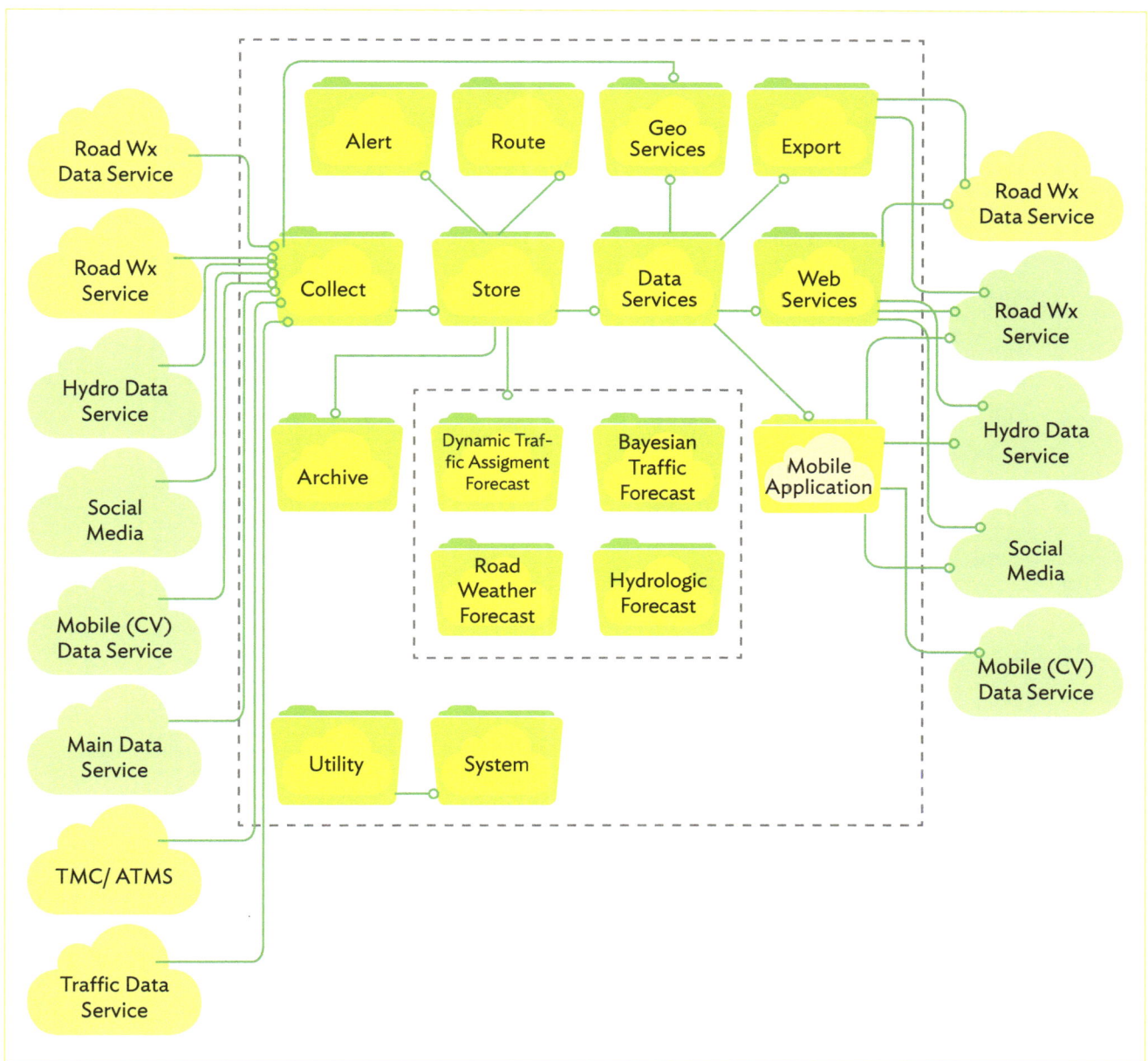

CV = connected vehicles, ATMS= advanced traffic management systems, TMS=traffic management systems, Wx = weather

Source: J.K. Garrett et al. 2017. Integrated Modeling for Road Condition Prediction (No. FHWA-JPO-18-631). United States. Dept. of Transportation. ITS Joint Program Office.

The diagram below in Figure 6 depicts the deployment of the IMRCP system. Items listed in the box on the left are deployed on the IMRCP server, and items listed on the right are being accessed by the IMRCP system through the internet.

Figure 6: Deployment of the Integrated Modeling for Road Condition Prediction System

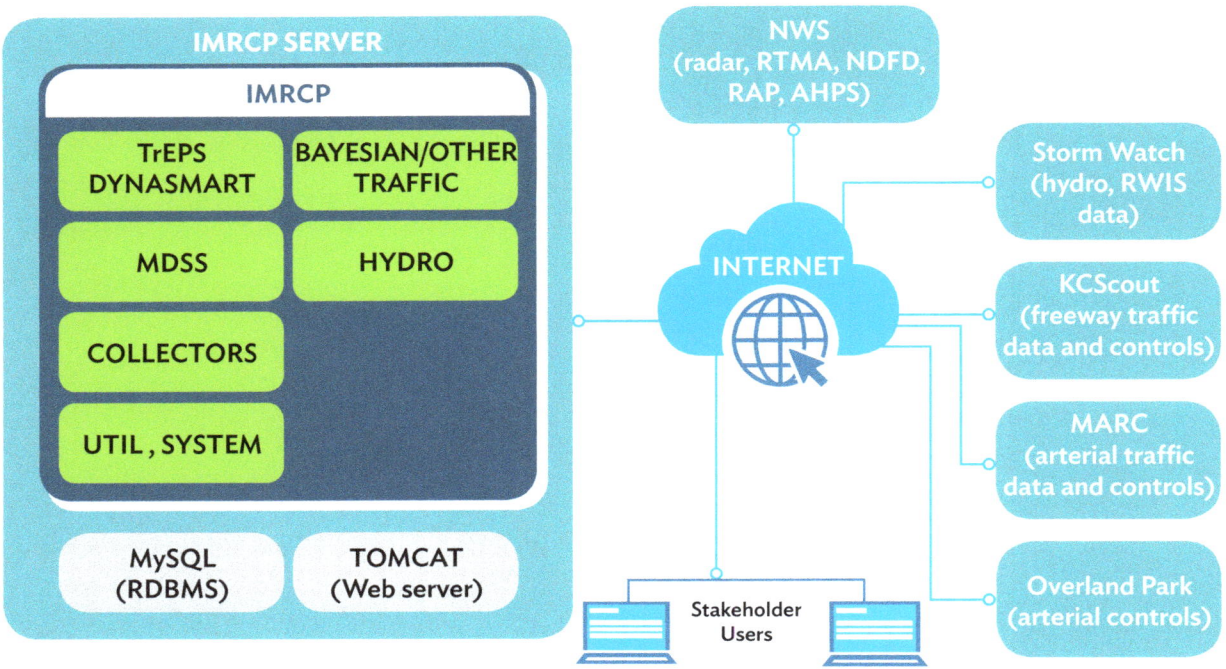

TrEPS=Traffic Estimation and Prediction System, NWS =National Weather Service, MDSS = Maintenance Decision Support System, HYDRO = Hydrological System, RWIS =road weather information system, RDBMS= Relational Database Management System, NDFD =National Digital Forecast Database, RTMA =Real-Time Model Assessment, RAP = Rapid Refresh AHPS = Advanced Hydrologic Prediction Service.

Source: J.K. Garrett et al. 2017. Integrated Modeling for Road Condition Prediction (No. FHWA-JPO-18-631). United States. Dept. of Transportation. ITS Joint Program Office.

The project is currently in phase 3–deployment and maintenance–the last steps of the SEP Vee diagram. The project has significantly benefited from the early stages of conceptual design by following the SEP. Based on the system design, KCScout has integrated it seamlessly into their existing traffic operations and maintenance activities. Based on the latest stakeholder engagement with KCScout, the IMRCP system has been a very effective subsystem addition to their existing management systems, and the functional objectives, as identified in the conceptual design, have been all successfully met. Based on the latest stakeholder engagement at a wider scale with other state transportation departments and traffic management centers, the IMRCP is well-received in those communities and additional deployments at other centers are being planned.

Chapter 3

Intelligent Transport Systems Conceptual Design —Gui'an New District Case Study

The ITS conceptual design (ITS CD) guides the development of an ITS suitable for financing. The ITS CD is part of the initial regional planning process in the systems engineering (SE) Vee diagram. The ITS CD development requires a series of steps summarized in Figure 7. The ITS CD has been undertaken with the support of relevant stakeholders that will develop, operate, maintain, or depend upon ITS. The system elements of the ITS CD represent relevant transportation management centers, field equipment (e.g., detectors, closed-circuit television or CCTV cameras, dynamic message signs, and weather stations), vehicles with ITS equipment (e.g., buses, commuter trains, and snow plows), and traveler equipment (e.g., mobile devices) that satisfy stakeholder needs. This chapter documents the methodologies and results of each of the six steps in Figure 4. For detailed information of the systems engineering analysis and stakeholder engagement, please refer to the project website.[4]

Figure 7: Steps for Developing the Intelligent Transport Systems Conceptual Design

STEP 1 Develop ITS objectives → **STEP 2** Identify stakeholders and establish the stakeholder working group → **STEP 3** Develop regional ITS architecture with customized service packages → **STEP 4** Allocate customized service packages to ITS Projects → **STEP 5** Validation by stakeholder working group → **STEP 6** Publish ITS conceptual design

ITS = intelligent transport systems.
Source: Asian Development Bank.

---

[4] Gui'an ITS Conceptual Design. http://www.consystec.com/guian/web/index.htm.

STEP 1: Develop ITS objectives.

Identification of ITS objectives is an important step to be able to later identify ITS services suitable for any city or region, in this case, Gui'an New District. ITS objectives should be based on the government's ITS plan.

Needs and Objectives

Objectives are statements of the impact that the ITS deployment will have on the non-ITS transportation infrastructure after deployment. Objectives should be specific and measurable so that they can be used to validate if the deployed ITS has the intended benefits.

The objectives for the Gui'an project were based on the development plans in the Gui'an New District,[5] and then were refined by conducting interviews with the members of the Stakeholder Working Group. Table 1 shows the objectives for the ITS in Gui'an. The objectives were used as the starting point for the selection and customization of services for the Gui'an Regional ITS Architecture.

Further, since there is currently virtually no ITS deployed in Gui'an and much of the future non-ITS transportation infrastructure is still to be deployed, the assumption is to estimate the improvement of operations or maintenance of the future transportation infrastructure with ITS deployed versus without ITS deployed.

In the end, 24 objectives were identified and organized into four objective groups (representing the overall needs categories): mobility (convenience and efficiency), safety, low carbon, and intelligence. The objectives are shown in the first column of the following four tables, one table for each objective group. The second column indicates the performance metric to measure and track the objective performance.

The key performance metrics can be influenced by more than one of the ITS services. For example, for the first mobility objective, "Improve (shorten) the travel time of public transit," this can be improved by off-board electronic payment (to shorten the bus dwell times at bus stops), by transit signal priority (to shorten the average time it takes a bus to travel through an intersection), or by automated bus lane enforcement (to shorten the time buses spend to travel on dedicated bus lanes by reducing the number of non-transit vehicles using the dedicated lanes). In most cases, the reported actual key performance indicators or KPI improvements were due to one specific ITS service, and do not take into consideration the improvements due to multiple ITS services that may have been deployed before or after the ITS improvement studied. To assure that the right objectives are identified and measured in a useful way, stakeholders should be engaged.

---

5   Bureau of Economic Development in Gui'an New District. 2018. Gui'an New District 13th Five Year Plan. Gui'an; and Guizhou Provincial Department of Housing and Urban Construction Department. 2014. Gui'an New District Masterplan (2013-2030). Guiyang.

Table 1: Objectives and Associated Metrics for the Intelligent Transport Systems in Gui'an New District

Objectives	Key Performance Metrics
Mobility (convenience and efficiency)	
Improve (shorten) the travel time of public transit.	Average minutes/trip
Improve the travel time reliability of public transit.	Standard deviation of trip time (minutes per trip)
Monitor and manage the occupancy of buses to avoid overcrowding.	Average number of passengers unable to sit on fixed-route public transit vehicles
	Average number of passengers unable to board fixed-route public transit vehicles
Make travel faster, more convenient, and a better experience (e.g., easy modal shifts).	Average passenger wait time per fixed route trip segments (minutes per segment)
Shorten commuting time in all modes.	Average time per trip
Prevent wide area congestion.	Percentage of trips delayed by stop-and-go travel (travel flow breakdown)
Reduce the travel delay comparable with similar sized cities.	Average minutes/trip
Residents feel satisfied with their travel experience.	Percentage survey respondents "satisfied" (versus "dissatisfied") with their travel experience
Increase the utilization of public transit.	Percentage of trips using public transit
Ensure high reliability (i.e., predictable travel times) of travel.	Standard deviation of actual trip time versus pre-trip predicted trip time (average minutes per trip)
Reduce roadway-recurring congestion and reduce congestion due to special events.	Percentage of trips impacted by recurring congestion percentage of trips impacted by special event congestion
Maintain roadways cost effectively and safely for travelers and maintenance workers.	Cost in RMB of roadway maintenance per month
	Number of injury crashes at roadworks per month
Preserve pavement from premature deterioration due to overweight vehicles.	Percentage of commercial vehicles that are detected as overweight
Safety	
Roads and intersections safety for all users is at the best levels for the PRC.	Average number of crashes per traveler km
Passenger transport safety is at the best levels for the PRC.	Average passenger injuries per km traveled

Table 1: continued

Objectives	Key Performance Metrics
Freight transport safety is at the best level for the PRC.	Average number of commercial vehicle crashes per commercial vehicle km traveled
Minimize illegal transport of passengers by detecting suspicious behavior.	Average number of passengers transported illegally per km of passenger trips
Low carbon	
Increase the number of trips that use green modes of travel e.g., public transit and non-motorized travel.	Ratio of (public transit trips + non-motorized travel trips) / total number of trips
Reduce carbon and hazardous pollutants due to transportation.	Average carbon pollution per trip Average particulate pollution per trip
Create a more livable travel environment.	Percentage of survey respondents "satisfied" (versus "dissatisfied") with the "livability" of their travel experience
Intelligence	
Provide convenient access to transport information services for road users.	Average time to plan a trip by users of trip planning services
Collect, archive, and analyze operational transportation data to improve future transportation investment planning.	Percentage of transportation planners "satisfied" (versus "dissatisfied") with the breadth, depth, availability of archived transportation operational data and standard reports using that data
Make it possible for people to experience future transport models and technologies.	Percentage of travelers "satisfied" (versus "dissatisfied") with the transport infrastructure and ITS technologies available in Gui'an
Manage efficient allocation of workers to their tasks.	Average percentage of the time for each transport worker allocated to productive work activities versus total work time

ITS = intelligent transport systems, PRC= People's Republic of China,. km = kilometer
Source: Asian Development Bank. 2019. *Gui'an New District ITS Conceptual Design*. Consultant's Draft Final Report. Manila (TA9437-PRC).

STEP 2: Identify stakeholders and establish the stakeholder working group.

One of the most important attributes of a successful ITS project is to gain the engagement of a group of senior individuals who represent each transportation mode or supporting services in a region to validate the project objectives and validate that "if we deploy an ITS consistent with the ITS CD, it will meet their needs." An ITS Working Group was formed for the project and it was made up of stakeholder representatives from major institutions that will operate and maintain ITS services in the Gui'an New District. The Gui'an ITS Working Group is made up of seven stakeholder institutional representatives covering all ITS services and modes, with one or two stakeholder representatives from each, summarized in Table 2. Table 2 identifies seven major ITS stakeholders ("Institution" column) and the working content or scope of their institution from an ITS perspective. All the stakeholder representatives were responsible for reviewing the Gui'an ITS Objectives early in the project, the ITS customized service packages (CSPs) that their ITS elements participated in, and the ITS CD analysis results toward the end of the project. The ITS stakeholder representatives need to critically review the ITS CD to make sure that, if a system is deployed to the requirements in the CD, that system would meet their needs.

Table 2: Gui'an Intelligent Transport Systems Working Group and their Scopes

	Institution	Working Content/Scope
1	Planning and Construction Office	Urban planning, urban construction industry management, urban road planning, streets (pedestrian crossing, pedestrian overpass), public parking lots
2	Law Enforcement Office	Urban roadway management, road maintenance
3	Cultural Tourism Investment Group Company	Public transportation operations, taxi operations
4	Public Transportation Office	Freight and commercial vehicles management, bus station management, bus dedicated lanes, emergency management
5	Police	Traffic safety enforcement
6	Cultural Tourism Center	Tourism industry management
7	Bureau of Economic Development Gui'an New District	All

ITS = intelligent transport systems.
Source: Asian Development Bank. 2019. *Gui'an New District ITS Conceptual Design*. Consultant's Draft Final Report. Manila (TA9437-PRC).

Developing measurable objectives using the key performance metrics for Gui'an had been difficult for the ITS Working Group. The solution was to develop a specific training for the stakeholders of the ITS Working Group in what formal objectives and KPIs were. The members of the working group were comfortable in selecting ITS technologies and services, but quantitative measurements were not easy for them to engage in discussion about.

Stakeholder Consultation on ITS conceptual design (photo by Robert S. Jaffe).

STEP 3: Develop regional ITS architecture with customized service packages.

The regional ITS architecture is central to conceptual design development. Once the objectives are in place, the framework ITS architecture is used as a "menu" of possible generic ITS services, and the selected ITS services from that menu are then customized based on the actual stakeholders in Gui'an and their actual stakeholder elements.

Framework Intelligent Transport Systems Architecture

The framework ITS architecture is based on generic ITS element subsystems as shown in Figure 8. In this diagram, the generic ITS element subsystems are organized in five groups:[6]

- iii. Center subsystems – can be located anywhere
- iv. Field subsystems – located on or near the roadway or right-of-way.
- v. Vehicle subsystems – located on or in a vehicle
- vi. Traveler subsystems – located with a traveler (like a smartphone) or a public information device (i.e., a "kiosk") that a traveler can approach to use
- vii. Support subsystems – usually located in a data center but present to support multiple elements.

---

6 Each of the subsystems in the groups is defined in more detail on the Architecture Reference for Cooperative and Intelligent Transportation (ARC-IT) version 8.1 website. The reference website for the framework ITS architecture is in https://local.iteris.com/arc-it/, and this is still being updated on a regular basis.

Figure 8: Generic Intelligent Transport Systems Elements in the Framework Architecture

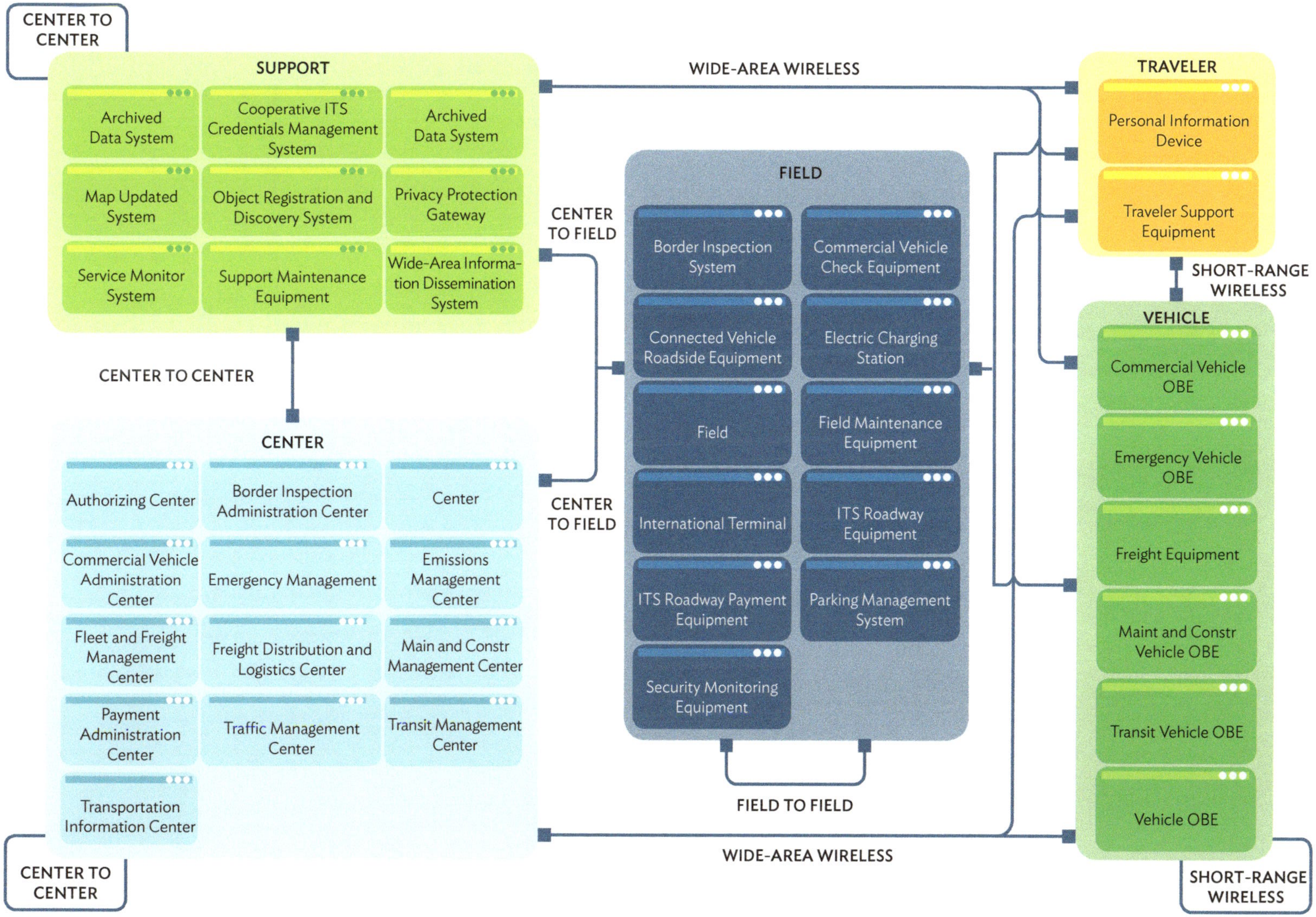

ITS = intelligent transport systems. OBE = on-board equipment
Source: US National ITS Architecture Version 8.1 (see https://www.its.dot.gov/research_archives/arch/index.htm)

Each of the subsystems has functional requirements defined for them, depending on in which service packages they are used. A subsystem can be used in many different service packages, and has different functional requirements based on each service package. The functional requirements of the subsystems are usually carried in the regional ITS architecture and used as a starting point for the detailed system requirements during the detailed design of the ITS.

In addition to generic subsystems, the framework ITS architecture also has elements called "terminators." Terminators are usually non-ITS elements with which the ITS interacts. Because they are non-ITS, they are not assigned functional requirements as they already exist, and the ITS need to adapt to the terminators, not the other way around.

The framework ITS architecture includes a set of over 100 generic ITS services covering all surface transportation modes.

Development of Customized Services Packages

After the Gui'an ITS objectives were developed and validated with the members of the ITS Working Group, the generic service packages were identified for customization. 72 generic service packages were (organized into 12 ITS service areas) identified. Service packages represent slices of the physical view that address specific services like traffic signal control. A service package collects together several different physical objects (systems and devices) and their functional objects and information flows that provide the desired service. The table below is a menu of service packages with underlying graphics and definitions.

Table 3: Intelligent Transport Systems Service Areas and their Service Packages in ARC-IT version 8.1

ITS Service Area	Generic Service Packages
Commercial vehicle (CV) operations	• Truck operations monitoring • Freight administration • Enforcement case cooperative handling • Cv administrative processes • Supervision of cv industry • Road weather information for freight carriers • Hazmat (hazardous materials) management • Detection of non-permitted dangerous/hazmat cargo • CV driver security authentication • Fleet and freight security
Data management	• ITS data warehouse • Performance monitoring

Table 3: continued

ITS Service Area	Generic Service Packages
Maintenance and construction (M&C)	• M&C vehicle and equipment tracking • Roadway automated treatment (anti-icing) • Winter maintenance • Roadway M&C • work zone management • work zone safety monitoring • M&C activity coordination
Parking management	• Parking space management • Parking electronic payment • Supervision of the parking industry • Regional parking management
Public safety	• Emergency call-taking and dispatch • Emergency vehicle preemption • Roadway service patrols • Evacuation and reentry management
Public transportation	• Taxi monitoring and dispatching • Transit vehicle tracking • Bicycle parking monitoring • Transit fixed-route operations • Public transportation industry supervision • Mobile enforcement management and supervision • Traffic enforcement data analysis and evaluation • Traffic enforcement information bulletin and service • Monitoring and review of bus incidents • Public transportation incident command and dispatch • Transit security • Transport hub operation monitoring and management • Taxi industry supervision • Transit passenger counting • Transit traveler information • Transit signal priority • Bus lane management • Multimodal coordination • Supervision of the bicycle-sharing industry • Integrated multimodal electronic payment

Table 3: continued

ITS Service Area	Generic Service Packages
Sustainable travel	• Roadside lighting • Electric charging stations management
Support	• Data distribution • Map management • Location and time • Privacy protection • Security and credentials management • Center maintenance
Traveler information	• Personalized traveler information • Infrastructure provided trip planning and route guidance • Information service at hub station
Traffic management	• Infrastructure-based traffic surveillance • Connected vehicle-based traffic surveillance • Traffic signal control • Connected vehicle traffic signal control • Traffic metering • Traffic information dissemination • Traffic incident management system • Integrated decision support and demand management • Traffic comprehensive monitoring • Dynamic roadway speed warning and enforcement
Vehicle safety	• Road weather motorist alert and warning
Weather	• Weather data collection • Weather information processing and distribution • Spot weather impact warning

ITS = intelligent transport systems.
Source: ARC-IT Service Packages. https://local.iteris.com/arc-it/html/servicepackages/servicepackages-areaspsort.html.

Each of the areas has one or more service packages to support one or several related services that fall into the "ITS Service Area." Some service packages can be viewed to support more than one area, but in this area, sorting each service package appears only once, and a decision was made to place the service package in one area or another. For example, the service package "Transit Signal Priority" is found in the public transportation area, but could just as well been placed in the traffic management area. Once a generic service package was selected for the Gui'an New District, it was added to a task list and its customization scheduled in the project plan.

The service package customization process involved deciding which stakeholders will be involved in a service, and replacing the generic ITS elements in ARC-IT with specific stakeholder elements in the RAD-IT.

The information flows between and among the stakeholder ITS elements in the customized service packages (CSPs) are developed in coordination with each stakeholder in the ITS Working Group. The information flows can be depicted in an example information flow diagram shown in Figure 9. The diagram shows the roll-up of information flows between the Cultural Tourism Investment (CTI) Enterprise Group Transit Vehicle On-Board Equipment (OBE), and the CTI Transit Management Center, based on multiple CSPs that each include a subset of the information flows.

Figure 9: Information Flow Diagram between Two Elements

CTI = Cultural Tourism Investment.
Source: Asian Development Bank. 2019. Gui'an New District ITS Conceptual Design, Consultant's Draft Final Report. Manila (TA9437-PRC).

In addition to information flow diagrams showing the information flows between two stakeholder ITS elements, a context diagram can be developed for each stakeholder element was developed. The context diagram for the CTI Transit Vehicle OBE, with the CTI Transit Vehicle OBE in the center and all other elements and all information flows, is shown in Figure 10.

Figure 10: Information Flow Context Diagram for Cultural Tourism Investment Transit Vehicle On-Board Equipment

Payment Device
- payment device updated
- request for payment
- payment
- payment device information

CTI Transit Management Center
- alarm notification
- fare collection data
- transit vehicle loading data
- transit vehicle location data
- transit vehicle schedule performance
- dispatch command-ud
- fare management information
- remote vehicle disable
- transit vehicle operator information
- transit schedule information

CTI Transit Vehicle Operator — **CTI Transit Vehicle On-Board Equipment**
- transit vehicle operator display
- transit vehicle operator input

----- Planned

CTI = Cultural Tourism Investment.
Source: Gui'an ITS Conceptual Design. http://www.consystec.com/guian/web/raditimages/ctx5.png.

The functional requirements for each CSP are also defined to be able to use the appropriate information flows. This is found in the list of ITS element functional requirements. In this case, there are functional requirements for each of the customized ITS services participated in by this element. For example, Transit Vehicle Passenger Counting yields the following functional requirements:

vii. The transit vehicle shall count passenger boarding and alighting;

viii. The passenger counts shall be related to location to support association of passenger counts with routes, route segments, or bus stops;

ix. The passenger counts shall be timestamped so that ridership can be measured by time of day and day of week; and

x. The transit vehicle shall send the collected passenger count information to the transit center.

These are just a subset of the many functional requirements allocated to the CTI Transit Vehicle OBE.

Customizing Service Packages

While the ITS service packages were customized for the objectives in Gui'an, in parallel the stakeholder inventory was developed as elements were needed to populate the CSPs. The final CSPs are shown on the project website under the "services" tab. Like the framework ITS architecture, the CSPs are organized by the same areas, with some of the framework service packages skipped as not being relevant to Gui'an New District stakeholders, and others having more than one instance to meet the various needs of the Gui'an New District. For example, there are three instances of the generic "Transit Vehicle Tracking" service package:

i. for tracking buses (in support of multiple services, including bus arrival traveler information);

ii. for tracking taxis (in support of more efficient taxi dispatching); and

iii. for tracking bicycles (to verify, and potentially to enforce, that bicycles are parked in allowable locations).

An example of a CSP for traffic management is shown in Figure 10. Note that the notation "TM01" is a short reference to the original generic service package in the framework ITS architecture. This CSP uses video-and detector-based surveillance of roadways to analyze congestion and support detection/classification/response of incidents by the Traffic Management Bureau's Traffic Management Center (TMB TMC or TMC, for short). The TMC shares road network conditions (e.g., link congestion and non-recurring incidents) with the Bureau of Economic Development in Gui'an New District (BEDGA) Public Transport Office (PTO) Transport Information Center (TIC). Each identified CSP has its own diagram and connecting flows like in Figure 11.

Figure 11: Customized Service Package for Basic Traffic Surveillance

**TM01: INFRASTRUCTURE-BASED TRAFFIC SURVEILLANCE
BASIC TRAFFIC SURVEILLANCE**

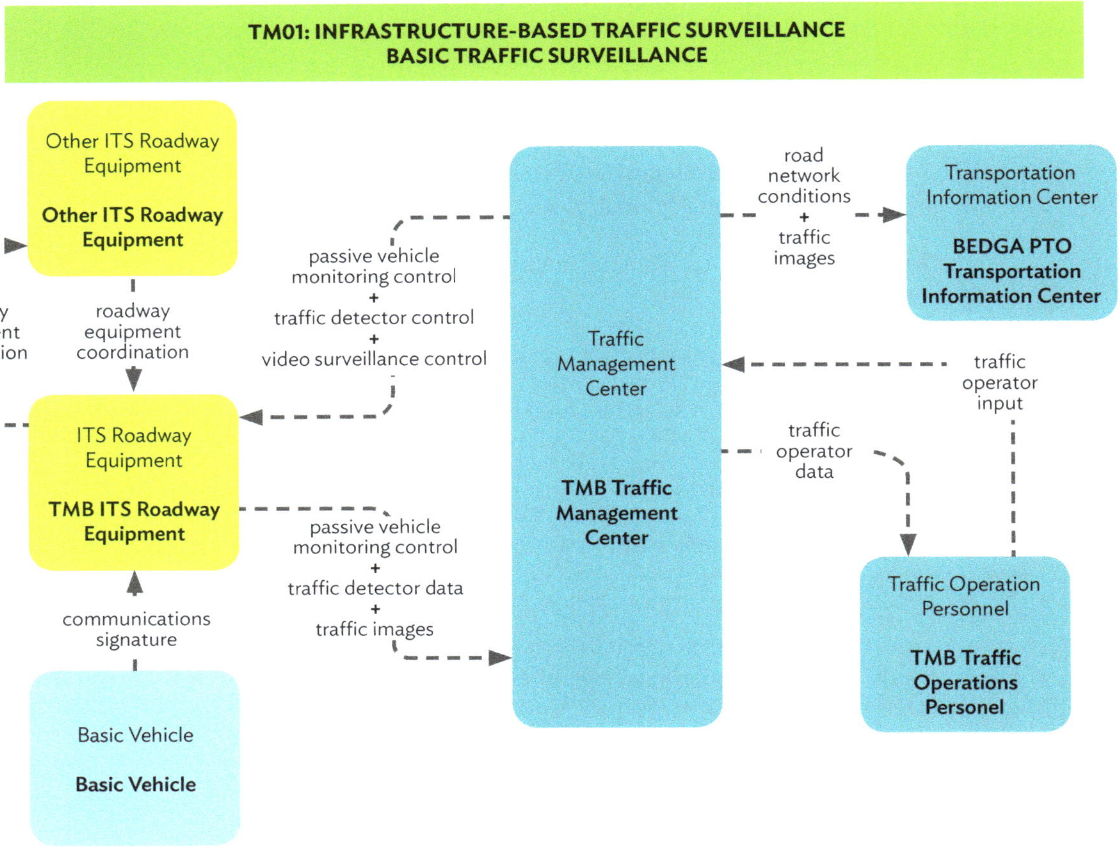

BEDGA = Bureau of Economic Development in Gui'an New District, ITS = intelligent transport systems, PTO = Public Transport Office, TMB = Traffic Management Bureau.
Source: Asian Development Bank. 2019. Gui'an New District ITS Conceptual Design, Consultant's Draft Final Report. Manila (TA9437-PRC).

Allocate customized service packages to ITS projects.

There are about 76 CSPs developed to meet all the Gui'an New District ITS objectives. Each CSP was allocated to an actual ITS project to be designed and then built. The nine projects, along with their major systems, are:

i. Real-time traffic and road-weather monitoring system

   (a) Road operation monitoring based on video

   (b) Full view video detection

   (c) Traffic flow detection

   (d) Information collection based on Internet of Things and electronic vehicle tags

   (e) Transportation security early warning based on cooperative vehicle–infrastructure system

   (f) Transportation weather and environment monitoring

ii. Big Data Service Center

iii. Multimodal Transportation Systems Management and Operations Center

    (a) Intelligent public transport management system

    (b) Integrated taxi management and service system

    (c) Intelligent parking management system

    (d) Intelligent bicycle management system

    (e) Dynamic operation supervision system for passenger vehicle (transit vehicle and taxi)

    (f) Dynamic operation supervision system for commercial vehicle

    (g) Integrated transportation hub information management system

    (h) Multimodal travel information service system

    (i) Traffic and transportation law enforcement management system

    (j) Maintenance and construction management system

iv. Electric Vehicle Application and Service System

v. Integrated Traffic Operations, and Security and Emergency Management System

    (a) Integrated traffic operation monitoring platform

    (b) Transportation security and emergency management platform

    (c) Decision support and information service platform

vi. Demonstration application of driverless vehicles and cooperative vehicle infrastructure system

    (a) Demonstration application of driverless vehicles

    (b) Demonstration application of cooperative vehicle infrastructure system

vii. Traffic Management Center

viii. Fleet and Freight Management Center

ix. Weather Management Center

Once the projects have been identified and "provisioned" with the CSPs, additional analysis can be done as needed. For example, in the case of Gui'an New District, the estimated cost, estimated benefits communication requirements, and other analyses for each project were prepared.

Verifying that all Objectives are Satisfied by the Customized Service Packages

It is essential that all objectives identified earlier are satisfied by the CSPs:

i. each objective must be satisfied by at least one CSP or a combination of CSPs; and

ii. each CSP must contribute to satisfying at least one objective (or a CSP which does not contribute to satisfying any objectives must be removed, or the set of objectives should be modified).

If an objective cannot be satisfied by one or several CSPs, then additional work needs to be done to either enhance one or more CSPs or create one or more new CSPs.

This verification was conducted by creating a traceability matrix of CSPs (the rows of the matrix) versus objectives (the columns of the matrix) (Table 4). If a CSP contributes to the satisfaction of an objective, then an "X" is placed in the cell where the CSP and the objective intersect.

Table 4: Objectives to Customized Service Packages Traceability Matrix for Gui'an New District

Count	Area	Short Name	Name	M1	M2	M3	M4	M5	M6	M7	M8	M9	M10	M11	M12	M13	S1	S2	S3	S4	L1	L2	L3	I1	I2	I3	I4
1		CVO02	Freight administration																								X
2		CVO04	CV administrative processes																								X
3		CVO10	Road weather information for freight carriers																					X			
4		CVO12	HAZMAT management																X								
5	Commercial vehicle operations	CVO13	Roadside HAZMAT security detection and mitigation																X								
6		CVO14	CV driver security authentication																X								X
7		CVO15	Fleet and freight security																X								
8		CVO08	Electronic work diaries																X							X	
9		CVO01	Truck operation monitoring																X							X	
10		CVO04	Supervision of commercial vehicle industry																								X
11	Data management	DM01	ITS Data Warehouse																							X	
12		DM02	Performance Monitoring																							X	

Table 4: continued

Customized Service Packages

Count	Area	Short Name	Name	Objectives																								
				Mobility													Safety				Low Carbon			Intelligence			
				M1	M2	M3	M4	M5	M6	M7	M8	M9	M10	M11	M12	M13	S1	S2	S3	S4	L1	L2	L3	I1	I2	I3	I4
13	Maintenance and construction	MC01	Maintenance and construction vehicle and equipment tracking												x												
14		MC03	Roadway automated treatment												x												
15		MC04	Winter maintenance												x												
16		MC05	Roadway maintenance and construction												x												
17		MC06	Work zone management												x												
18		MC07	Work zone safety monitoring												x												
19		MC08	Maintenance and construction activity coordination				x							x	x												
20	Parking management	PM01	Parking space management				x				x																
21		PM03	Parking electronic payment				x				x																
22		PM04	Regional parking management				x				x																
23		PM03	The supervision of the parking industry																								x

Table 4: *continued*

Customized Service Packages				Objectives																							
				Mobility													Safety				Low Carbon			Intelligence			
Count	Area	Short Name	Name	M1	M2	M3	M4	M5	M6	M7	M8	M9	M10	M11	M12	M13	S1	S2	S3	S4	L1	L2	L3	I1	I2	I3	I4
24	Public safety	PS01	Emergency call-taking and dispatch														X										
25		PS03	Emergency vehicle preemption														X										
26		PS08	Roadway service patrols								X																
27		PS13	Evacuation and reentry management								X						X										
28		PT02	Transit fixed-route operations	X												X											
29	Traffic enforcement	PT04	Enforcement cooperative case handling																	X							
30		PT04	Mobile enforcement management and supervision																	X							X
31		PT04	Traffic enforcement data analysis and evaluation																							X	
32		PT04	Traffic enforcement information bulletin and service																						X		

Table 4: continued

Customized Service Packages

Count	Area	Short Name	Name	Objectives														Safety				Low Carbon				Intelligence			
				M1	M2	M3	M4	M5	M6	M7	M8	M9	M10	M11	M12	M13	S1	S2	S3	S4	L1	L2	L3	I1	I2	I3	I4
33	Public transportation	PT01	Transit vehicle tracking		X													X								X		
34		PT05	Transit security															X										
35		PT07	Transit passenger counting			X	X				X																	
36		PT08	Transit traveler information								X														X			
37		PT09	Transit signal priority	X	X																							
38		PT14	Multimodal coordination					X	X	X			X															
39		PT04	Public transportation industry supervision																									X
40		PT05	Monitoring and review of bus incidents															X										
41		PT10	Bus lane management	X	X																							
42		PT05	Public transportation incident command and dispatch															X										
43		PT07	Public transportation network optimization and decision support	X				X	X		X	X														X		
44		PT01	Taxi monitoring and dispatching					X			X	X																
45		PT06	Taxi industry supervision																									X
46		PT05	Transport hub operation monitoring and management								X																	
47		PT18	Network ticketing	X			X				X																	

Table 4: *continued*

Customized Service Packages

Count	Area	Short Name	Name	Mobility M1	M2	M3	M4	M5	M6	M7	M8	M9	M10	M11	M12	M13	Safety S1	S2	S3	S4	Low Carbon L1	L2	L3	Intelligence I1	I2	I3	I4
48	Support	SU03	Data distribution																						X		
49		SU04	Map management																					X		X	
50		SU05	Location and time																					X	X	X	
51		SU07	Privacy protection																					X			
52		SU08	Security and credentials management																					X		X	
53		SU09	Center maintenance																					X			
54		ST04	Roadside lighting																		X	X					
55		ST05	Electric charging stations management					X				X												X	X		
56	Sustainable travel	ST05	New-energy vehicle integrated management								X														X		
57		PT14	Supervision of bicycle sharing industry					X				X									X						
58		PT01	Bicycle parking monitoring			X																					X

Table 4: *continued*

Customized Service Packages

Count	Area	Short Name	Name	Objectives - Mobility													Safety				Low Carbon			Intelligence			
				M1	M2	M3	M4	M5	M6	M7	M8	M9	M10	M11	M12	M13	S1	S2	S3	S4	L1	L2	L3	I1	I2	I3	I4
59		TM01	Infrastructure-based traffic surveillance																							x		
60		TM02	Vehicle-based traffic surveillance																							x		
61		TM03	Traffic signal control				x	x	x	x																		
62		TM04	Connected vehicle traffic signal system							x																	x	
63		TM05	Traffic metering																						x	x		
64	Traffic management	TM06	Traffic information dissemination																						x			
65		TM08	Traffic incident management system														x											
66		TM09	Integrated decision support and demand management																							x		
67		TM12	Dynamic roadway warning														x											
68		TM17	Speed warning and enforcement														x											
69		TM09	Traffic comprehensive monitoring																						x	x		

Table 4: continued

Customized Service Packages			Objectives																									
			Mobility													Safety				Low Carbon			Intelligence				
Count	Area	Short Name	Name	M1	M2	M3	M4	M5	M6	M7	M8	M9	M10	M11	M12	M13	S1	S2	S3	S4	L1	L2	L3	I1	I2	I3	I4
70	Traveler information	TI02	Personalized traveler information																					X			
71		TI04	Infrastructure-provided trip planning and route guidance																					X			
72		TI05	Information service at hub station																					X			
73	Vehicle safety	VS07	Road weather motorist alert and warning														X							X			
74	Weather	WX01	Weather data collection																						X		
75		WX02	Weather information processing and distribution																					X			
76		WX03	Spot weather impact warning														X							X			

HAZMAT = Hazardous Materials, ITS = intelligent transport systems.

Source: Asian Development Bank. 2019. *Gui'an New District ITS Conceptual Design*. Consultant's Draft Final Report. Manila (TA 9437-PRC).

STEP 5: Stakeholder validates the customized service packages.

After the initial draft of the complete set of CSPs were prepared, the members of the Working Group in a series of individual meetings reviewed each of the CSPs participated in by their ITS elements. The Working Group members also at that time revisited their needs, objectives, key performance indicators, and scope. Each CSP was also deliberated in detail with each Working Group member. These meetings took from a few hours to multiple days in some cases. The meetings were a combination of training for the Working Group members to explain the ITS services and the full scope of the possible ITS services that their function might participate in or lead, and then editing the draft CSPs to remove and add functionality as the Working Group members thought was best for their function in the Gui'an New District.

An example of the change in a CSP after validation by the working group member was the CSPs in the CSP groups of Data Management, Parking Management, and Sustainable Travel. After reviewing all the CSPs for the ITS Data Warehouse function, the Working Group stakeholder asked if the system was going to archive the actual routes taken by the commercial vehicles (CVs). It turns out that only credentials and permits for each CV were being stored, and not the actual routes taken, which was available in the "CTI Fleet and Freight Management Center." Based on the stakeholder's comment, the CSP, which identifies the CV sources of data for the Data Archive as shown in Figure 12, was updated.

Figure 12: Intelligent Transport Systems Data Warehouse

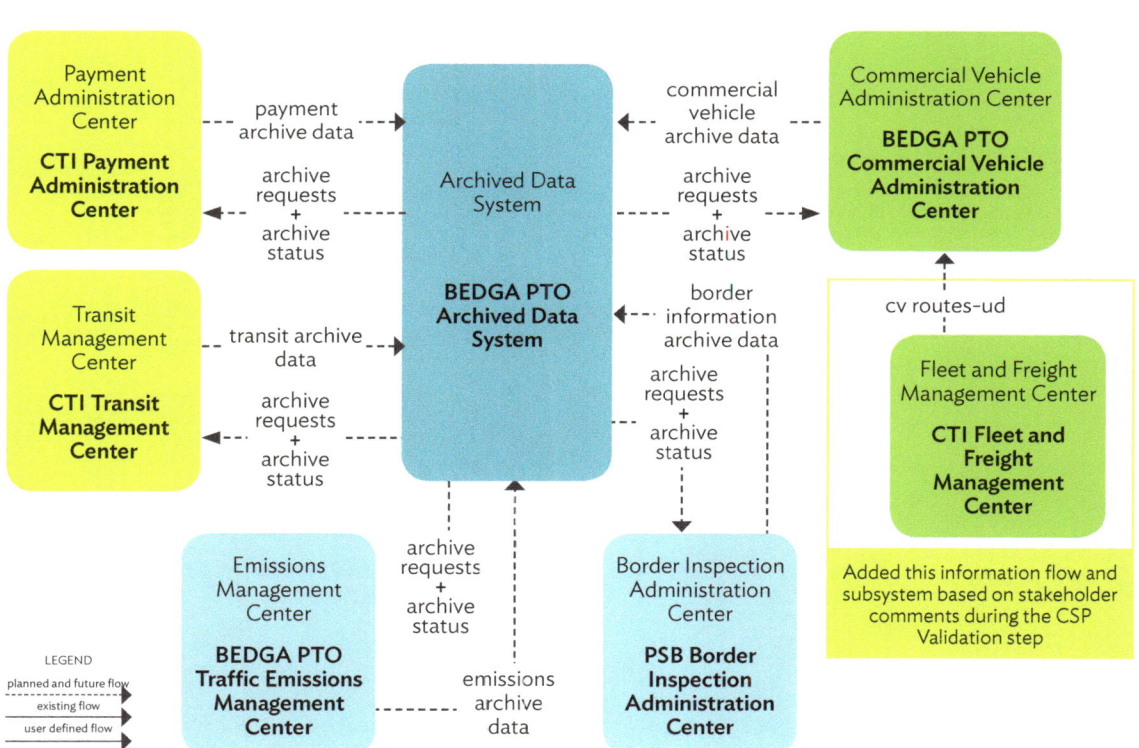

BEDGA = Bureau of Economic Development in Gui'an New District, CSP = customized service package, CTI = Cultural Tourism Investment, ITS = intelligent transport systems, PSB = Public Security Bureau, PTO = Public Transport Office.

Source: Asian Development Bank. 2019. Gui'an New District ITS Conceptual Design, Consultant's Draft Final Report. Manila (TA9437-PRC).

The boxed center and information flow on the right side of the diagram were added to this specific ITS Data Warehouse information flow diagram, which specifies that "CV Routes," a new (user-defined) information flow, will be sent to the BEDGA PTO Commercial Vehicle Administration Center, and the information about actual vehicle routes taken can be included in the existing "commercial vehicle archive data" flow.

STEP 6: Publish ITS conceptual design.

Because the ITS CD has a very large amount of information that will be used by the developers of the formal ITS Feasibility Study Report (FSR), which will include the detailed design requirements for the ITS projects, select technologies, and prepare detailed specifications, it is recommended that the ITS CD be published on the internet using a highly cross-referenced methodology. For example see, the website for the Gui'an New District ITS CD: http://www.consystec.com/guian/web/index.htm.

Figure 13: Gui'an Intelligent Transport Systems Conceptual Design

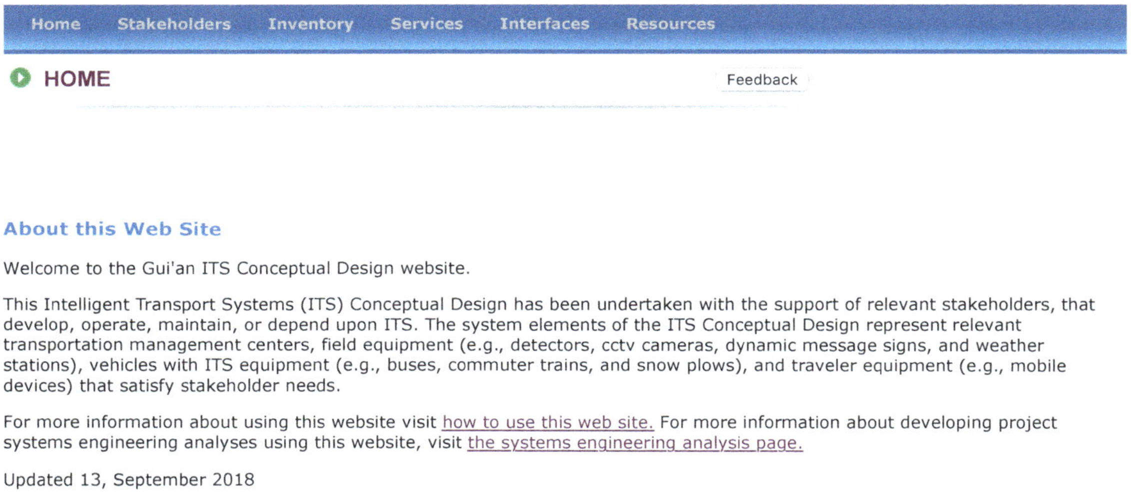

Home Stakeholders Inventory Services Interfaces Resources

○ **HOME**   Feedback

About this Web Site

Welcome to the Gui'an ITS Conceptual Design website.

This Intelligent Transport Systems (ITS) Conceptual Design has been undertaken with the support of relevant stakeholders, that develop, operate, maintain, or depend upon ITS. The system elements of the ITS Conceptual Design represent relevant transportation management centers, field equipment (e.g., detectors, cctv cameras, dynamic message signs, and weather stations), vehicles with ITS equipment (e.g., buses, commuter trains, and snow plows), and traveler equipment (e.g., mobile devices) that satisfy stakeholder needs.

For more information about using this website visit how to use this web site. For more information about developing project systems engineering analyses using this website, visit the systems engineering analysis page.

Updated 13, September 2018

Source: Gui'an ITS Conceptual Design. http://www.consystec.com/guian/web/index.htm.

Chapter 4

Intelligent Transport Systems Project Planning and Recommendations

This chapter lays out the activities, resources, budget, and time line for the component projects of the Gui'an New District Intelligent Transport Systems (ITS).

Intelligent Transport Systems Conceptual Design Project Schedule

The development of the ITS conceptual design (ITS CD) required about 7 months. Figure 14 shows the time line for the Gui'an New District conceptual design development. An international consultant (80 workdays) and two national consultants were engaged to develop the ITS CD. The work included training/mentoring the national consultants to support the ITS CD development using the Architecture Reference for Cooperative and Intelligent Transportation (ARC-IT) framework and the Regional Architecture Development for Intelligent Transport (RAD-IT) development tool.

Figure 14: Gantt Chart for the Gui'an New District Conceptual Design Development

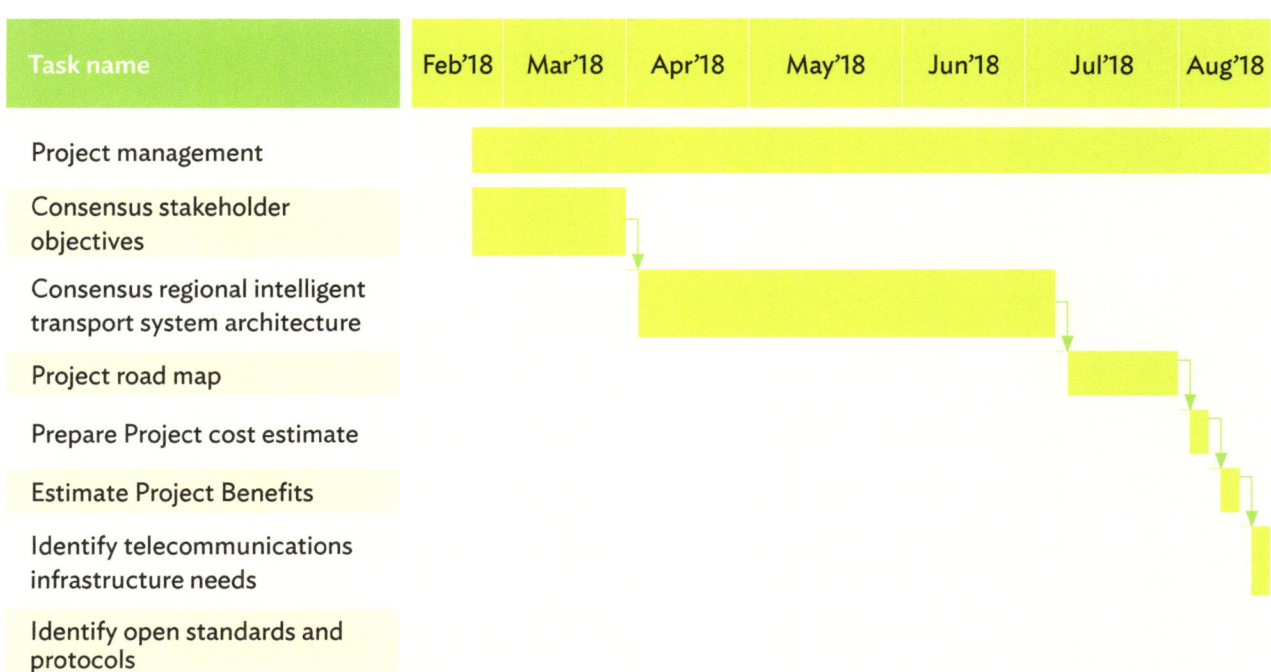

Source: Asian Development Bank. 2019. *Gui'an New District ITS Conceptual Design*. Consultant's Draft Final Report. Manila (TA9437-PRC).

Estimated Intelligent Transport Systems Project Design-and-Build Schedule

The estimated schedule for each of the ITS projects and their major subsystems is shown in Table 5: The estimated schedule identifies four phases for each of the project subsystems:

i. **Recruit detailed design institute.** The detailed design will be done by consultants from a design institute, and the recruitment of the design institute will be managed by the Bureau of Economic Development in Gui'an New District (BEDGA) project management office (PMO). The selection scoring of proposing design institutes will be based on their understanding of, and capabilities to conduct, the detailed design required for the systems in the ITS Feasibility Study Report (FSR). The ITS FSR should include or reference all the requirements of the ITS CD, including the Gui'an Regional ITS Architecture and the associated interface open standards.

ii. **Detailed design.** This will be done by the selected design institute identified/recruited in the prior phase. The Stakeholder Working Group assembled for the ITS CD should be enlisted to review the intermediate and detailed design deliverables to validate that, if the ITS are built based on the detailed designs, then it will meet their needs.

iii. **Procurement of works.** A separate procurement for the build and installation, and testing contractor for the ITS will be managed by the BEDGA PMO. The detailed design of the prior phase will be the basis for scoring the construction capabilities of the bidding contractors.

iv. **Build and Installation (begin *operation and maintenance* phase on completion).** After the systems are built and tested per the detailed design specifications, which should include operations and maintenance documentation and staff training, the systems will be released to operations and maintenance by BEDGA or associated stakeholders. At this point, the operational systems should be validated that the original objectives have been achieved.

4. IIntelligent Transport Systems Project Planning and Recommendations

Table 5: Schedule for the Project Subsystems

Project Serial No.	Project Name	System Serial No.	System Name	Start Year of Procurement	Start Year of Operation	Construction Period (from draft ITS FSR)					
						Year 1	Year 2	Year 3	Year 4	Year 5	Year 6
						Recruit Detailed Design Institute	Detailed Design	Procurement of Works	Build/Installation (Begin Operations on completion)		
One	Real-time traffic and road-weather monitoring system	1	Road operation monitoring based on video	Year 1	Year 4						
		2	Full view video detection	Year 1	Year 4						
		3	Traffic flow detection	Year 1	Year 4						
		4	Information collection based on Internet of Things and vehicle electronic tags	Year 2	Year 6						
		5	Transportation security early warning based on cooperative vehicle infrastructure system	Year 2	Year 6						
		6	Transportation weather and environment monitoring	Year 1	Year 4						
Two	Big Data Service Center	7	Smart transportation big data center	Year 2	Year 6						
Three	Multimodal Transportation Systems Management and Operations Center	8	Intelligent public transport management system	Year 1	Year 4						
		9	Intelligent taxi management and service system	Year 1	Year 4						
		10	Intelligent parking management system	Year 1	Year 4						
		11	Intelligent bicycle (shared bicycle) management system	Year 1	Year 4						
		12	Dynamic operation supervision system for passenger vehicle (transit vehicle and taxi)	Year 1	Year 4						
		13	Dynamic operation supervision system for commercial vehicle	Year 1	Year 4						
		14	Integrated transportation hub information management system	Year 1	Year 5						
		15	Multimodal travel information service system	Year 1	Year 5						
		16	Traffic and transportation law enforcement management system	Year 2	Year 6						
		17	Maintenance and construction management system	Year 1	Year 5						
Four	Electric Vehicle Application and Service System	18	New-energy vehicle application and service system	Year 1	Year 5						
Five	Integrated Traffic Operations, and Security and Emergency Management System	19	Integrated traffic operation monitoring platform	Year 3	Year 6						
		20	Transportation security and emergency management paltform	Year 3	Year 6						

ITS FSR = Intelligent Transport S Systems Feasibility Study Report.
Source: Asian Development Bank. 2019. *Gui'an New District ITS Conceptual Design.* Consultant's Draft Final Report. Manila (TA9437-PRC).

Recommended Resources and Specialty Skills for Detailed Design

Special resources and specialties needed by the selected design institutes for the detailed design of the ADB-funded projects are summarized in Table 6.

Table 6: Special Resources and Specialties Recommended for the Detailed Design of the Projects

Project Number	Project Name	System Serial Number	System Name	Recommended Resources and Specialties Needed for Detailed Design
One	Real-time Traffic and Road-weather Monitoring System	1	Road operation monitoring based on video	Intelligent Transport Systems Civil Engineering Traffic Engineering Electrical Engineering Environmental Engineering Communication Engineering Cooperative Vehicle Infrastructure System
		2	Full view video detection	
		3	Traffic flow detection	
		4	Information collection based on Internet of Things and vehicle electronic tag	
		5	Transportation security early warning based on cooperative vehicle infrastructure system	
		6	Transportation weather and environment monitoring	
Two	Big Data Service Center	7	Smart transportation big data center	Intelligent Transportation System(ITS) Systems Analysis and Integration Communication Engineering Computer Science Data Science

Table 6: *continued*

Project Number	Project Name	System Serial Number	System Name	Recommended Resources and Specialties Needed for Detailed Design
Three	Multimodal Transportation Systems Management and Operations Center	8	Intelligent public transport management system	Intelligent Transportation System(ITS) Systems Analysis and Integration Communication Engineering Systems Engineering Software Engineering
		9	Integrated taxi management and service system	Intelligent Transportation System(ITS) Systems Analysis and Integration Communication Engineering Systems Engineering Software Engineering
		10	Intelligent parking management system	
		11	Intelligent bicycle (shared bicycle) management system	
		12	Dynamic operation supervision system for passenger vehicle (transit vehicle and taxi)	
		13	Dynamic operation supervision system for commercial vehicle	
		14	Integrated information hub information management system	
		15	Multimodal travel information service system	
		16	Traffic and transportation law enforcement management system	
		17	Maintenance and construction management system	

Table 6: *continued*

Project Number	Project Name	System Serial Number	System Name	Recommended Resources and Specialties Needed for Detailed Design
Four	Electric Vehicle Application and System	18	New-energy vehicle application and service system	System Analysis and Integration Communication Engineering Environmental Engineering Vehicle Engineering Software Engineering
Five	Integrated Traffic Operations, and Security and Emergency Management System	19	Integrated traffic operation monitoring platform	Intelligent Transport Systems Analysis and Integration Traffic Engineering Software Engineering
		20	Transportation security and emergency management platform	
		21	Decision support and information service platform	

Source: Asian Development Bank. 2019. *Gui'an New District ITS Conceptual Design*. Consultant's Draft Final Report. Manila (TA9437-PRC).

Project Analysis

The project cost was the derived for the ADB funded project components which costs about $93 million. This includes the design and build and assembly of the systems.

The ITS project viability was assessed by estimating the ITS project benefits. Quantitative benefits are generally difficult to estimate for Gui'an because currently there are no available traffic studies or orgin-desination studies for traveler demand or freight movements demand. Also, traveler preferences are unknown for transit versus mobility on demand (MOD) (e.g., DiDi) versus private vehicles versus nonmotorized transport (bicycles and walking). Despite this lack of empirical data on which to base benefits for the deployment of ITS services in Gui'an, quantitative benefits from other areas were collected. Some of the benefits of the ITS project were reduction in travel time, safety improvement, emissions reduction and fuel consumption savings. For more detailed project benefits please refer to Appendix 3.

Communications requirements for the system were also assessed. A combination of three communications technologies were determined to be deployed in Gui'an. For more information on the communications requirements, see Appendix 4.

There are several open standards and protocols the will determine interoperability, interface compatibility of different elements. These standards are discussed in Appendix 5.

Recommendations

Developing ITS benefits. ITS benefits should be measured once the ITS are in place to be able to evaluate the actual benefits of the system. A before and after study is recommended and should be conducted to develop many of the efficiency benefits of ITS investments. The performance of the transportation system should be measured before and after deployment of the ITS components. To do this, a minimum origin-destination travel demand analysis and the corresponding traffic model analysis should be conducted. For safety benefits, the typical safety problems in similar areas of the PRC should be known (e.g., number of deaths and injuries per kilometer traveled or per trip on each mode).

Project implementation. The following specific actions should be taken for the detailed design of an ITS project:

i. Include a traceability analysis so that it can be verified that all ITS services are covered by the technology selections, and each technology selection supports at least one ITS service.

ii. Use the identified customized service packages as requirements for the detailed design.

iii. Perform similar tests on the completed systems: that they exchange information and implement services as specified in the conceptual design.

iv. Hire an independent verification and validation consultant to verify the detailed design and validate that the constructed system meet the above requirements derived from the ITS conceptual design. The detailed design of the project should be based on the ITS CD developed.

Chapter 5

Conclusions

Under Strategy 2030 of the Asian Development Bank (ADB), sharing of experiences, best practices, and innovations is becoming increasingly important in engagement, especially in critical areas such as regional public goods, climate change, urbanization, and regional cooperation initiatives as well as emerging social issues, such as aging. ADB proactively seeks ways to promote the use of advanced technologies across its operations and provide capacity-building support to developing member countries. ADB also promotes mainstreaming of advanced technologies in infrastructure operations that reduce life cycle costs and increase durability, improve the efficiency, and maintain the quality of services, while minimizing negative environmental and social impacts.

Gui'an New District is planned to be the most dynamic growth pole in Guizhou Province and southwest People's Republic of China (PRC). It will be built as a safe, orderly, smooth, accessible, low-carbon, and sustainable smart city. Among the ways to achieve the objectives include the (i) development of intelligent transport systems (ITS) that incorporate advanced information processing technology, data communication technology, electronic sensing technology, control technology, Internet of Things, and cloud computing methods in a big data center; (ii) promotion of environment-friendly public transport with the use battery electric buses to support mobility; and (iii) advancement of the pilot study on the intelligent connected vehicles system for an emerging industry that makes use of the information and communication technology revolution. These high-technology features and facilities will be made more inclusive using human-centered design.

The systems engineering approach to developing and implementing high-technology projects is a good tool to use and understand because it focuses on managing the risks in the design, deployment, and maintenance of complex interacting elements over their life cycles. The bulk of this report discusses the intelligent transport systems conceptual design (ITS CD), which is part of the feasibility study or concept exploration phase of the systems engineering process. This is essential as it fully illustrates the needs of stakeholders, and plans and designs the ITS based on their objectives. This is also an important part of the whole process—from planning to implementation of an ITS project.

A stakeholder needs and objective-based ITS was developed for the Gui'an New District. The conceptual design includes a regional architecture that shows the stakeholder elements and their information-sharing interfaces with other elements. It is intended to ensure that implementation of ITS services meets all the stakeholder objectives. The ITS architecture in the conceptual design is a technology-neutral representation of the ITS services, and it should be used as information-sharing requirements to guide the detailed design, construction, and testing of the ITS elements. The detailed conceptual design is fully documented in http://www.consystec.com/guian/web/index.htm.

The development of ITS CD in a project can enable future system operators, maintainers, and users to plan ITS services based on existing best practices that meet the specific transport safety and efficiency needs of Gui'an New District. Its development at the very beginning of a regional transportation technology project is intended to avoid very costly interoperability (or institutional integration) defects later in project procurement, operation, and maintenance.

To develop a comprehensive and sustainable ITS, it is important to invest in a regional architecture or ITS CD. Once this is developed and identified, city or municipal governments can ensure that ITS projects they finance will be able to adapt to changes, such as frequent equipment and technology advancements.

Encouraging the development of a proper regional architecture or ITS CD is also important for development partners like ADB before investing in an ITS project to avoid costly and piecemeal ITS investments. It will help ensure investment returns and provide knowledge sharing. In Asia and the Pacific, ADB should further support the development of regional architecture like cooperative ITS that requires crossborder collaboration, including institutional and legislative agreements. This also strengthens ADB's responsibility and role as an honest broker of regional prosperity and compatibility in the technological front—as a robust ITS can promote sustainable growth, improve transport efficiency, and significantly ease traffic flow.

References

Garrett, J.K. et al. 2017. Integrated Modeling for Road Condition Prediction (No. FHWA-JPO-18-631). United States. Dept. of Transportation. ITS Joint Program Office.

Sage, A. P. 1992. *Systems engineering* (Vol. 6). John Wiley & Sons.

US Department of Transportation. *Connected Vehicles. Intelligent Transportation Systems Joint Program Office.* https://www.its.dot.gov/cv_basics/index.htm.

US Department of Transportation. 2007. *Systems Engineering for Intelligent Transportation Systems: An Introduction for Transportation Professionals. Federal Highway Administration. Federal Transit Organization.* https://ops.fhwa.dot.gov/publications/seitsguide/seguide.pdf.

US Department of Transportation. *Architecture Reference for Cooperation and Intelligent Transportation.* https://local.iteris.com/arc-it/.

Appendix 1

List of Stakeholders and Their Intelligent Transport Systems Elements

Bureau of Economic Development in Gui'an New District (BEDGA)
Public Transportation Office (PTO)

- BEDGA PTO Alerting and Advisory System
- BEDGA PTO Alternate Mode Transportation Center
- BEDGA PTO Archived Data Administrator
- BEDGA PTO Archived Data System
- BEDGA PTO Asset Management System
- BEDGA PTO Commercial Vehicle Administration Center
- BEDGA PTO Commercial Vehicle Administrator
- BEDGA PTO Commercial Vehicle Check Equipment
- BEDGA PTO Connected Vehicle Roadside Equipment
- BEDGA PTO Cooperative Intelligent Transport Systems (ITS) Credentials Management System
- BEDGA PTO Credentials Management System Operator
- BEDGA PTO Data Distribution System Operator
- BEDGA PTO Data Distribution Systems
- BEDGA PTO Emergency Management Center
- BEDGA PTO Enforcement Management Personnel
- BEDGA PTO Event Promoter System
- BEDGA PTO Map GIS Operator
- BEDGA PTO Map Update System
- BEDGA PTO Mobile Enforcement Device
- BEDGA PTO Network Time Source
- BEDGA PTO New-Energy Vehicle Management Center
- BEDGA PTO Object Registration and Discovery System
- BEDGA PTO Privacy Protection Gateway

- BEDGA PTO Security Monitoring Equipment
- BEDGA PTO Service Monitoring System
- BEDGA PTO Service Monitoring System Operator
- BEDGA PTO Surface Transportation Weather Service
- BEDGA PTO Transport Information Center Operator
- BEDGA PTO Traffic Emissions Management Center
- BEDGA PTO Traffic Enforcement Center (Transportation)
- BEDGA PTO Traffic Operation Monitoring Center
- BEDGA PTO Transit Management Personnel
- BEDGA PTO Transportation Information Center
- BEDGA PTO Travel Services Provider System
- BEDGA PTO Wide Area Information Disseminator System

City Management Bureau (CMB)

- CMB Asset Management System
- CMB Basic Maintenance and Construction Vehicle
- CMB Field Maintenance Equipment
- CMB ITS Roadway Equipment
- CMB Maintenance and Construction Administrative System
- CMB Maintenance and Construction Center Personnel
- CMB Maintenance and Construction Field Personnel
- CMB Maintenance and Construction Management Center
- CMB Maintenance and Construction Vehicle On-Board Equipment (OBE)
- CMB Maintenance and Construction Vehicle Operator
- CMB Storage Facility Data Acquisition System

Cultural Tourism Investment Enterprise Group (CTI)

- CTI Basic Commercial Vehicle
- CTI Commercial Vehicle Driver
- CTI Commercial Vehicle OBE
- CTI Financial Center
- CTI Fleet and Freight Management Center
- CTI Fleet-Freight Manager
- CTI Freight Distribution and Logistics Center
- CTI Payment Administration Center
- CTI Taxi Management Center
- CTI Taxi OBE
- CTI Taxi Vehicle

- CTI Transit Management Center
- CTI Transit Operations Personnel
- CTI Transit Vehicle OBE
- CTI Transit Vehicle Operator
- CTI Vehicle Location and Time Data Source

Emergency Management Office (EMO)

- EMO Emergency Management Center

Finance Bureau

- Finance Bureau Financial Center

Other Regions

- Other Archived Data Systems
- Other Credentials Management Systems
- Other Commercial Vehicles Administration Centers
- Other Data Distribution System
- Other Emergency Management Centers
- Other ITS Roadway Equipment
- Other Maintenance and Construction Mangement Centers
- Other Map Update Systems
- Other Maintenance and Construction Vehicle OBEs
- Other Payment Administrator Centers
- Other Private Company Parking Management Systems
- Other Traffic Enforcement Center
- Other Traffic Management Centers
- Other Transit Management Centers
- Other Transportation Information Centers

Planning and Construction Bureau

Note: This is a key user of the ITS, but does not have any ITS elements in the inventory.

Private Company

- Private Company Parking Management System
- Private Company Parking Operator

Public Security Bureau (PSB)

- PSB Border Inspection Administration Center
- PSB Border Inspection System

- PSB Comprehensive Application Platform of Traffic Management (6 in 1)
- PSB Emergency Telecommunications System
- PSB EMO Emergency Personnel
- PSB EMO Emergency System Operator
- PSB EMO Emergency Vehicle OBE

Shared Bicycle Operation Company (SBOC)

- SBOC Shared Bicycle
- SBOC Shared Bicycle Operation Center

Traffic Management Bureau (TMB)

- TMB ITS Roadway Equipment
- TMB Traffic Management Center
- TMB Traffic Operations Personnel

Travelers

- Basic Vehicle
- Cyclist
- Driver
- Pedestrian
- Personal Information Device
- Traveler
- Traveler Support Equipment
- Vehicle OBE

Appendix 2

Customized Services Packages for Gui'an New District

Appendix 2 aids understanding the customized service packages (CSPs) in the intelligent transport systems conceptual design (ITS CD), and in particular, the CSPs as documented on the ITS CD website, http://www.consystec.com/guian/web/services.htm.

Table A2: Customized Services Packages for Gui'an New District

ITS Service Area	Service Packages	Functionality
Commercial vehicles (CV) operations	Truck Operations Monitoring (CVO01)	The Cultural Tourism Investment Enterprise Group (CTI) Fleet and Freight Management Center monitors the CTI commercial vehicles via their on-board equipment (OBE). Monitoring consists of detecting breaches of freight containers; reading a vehicle's carrier ID; driver ID; and vehicle ID, including the status of freight equipment; reading onboard safety/vehicle data; and reading the vehicle location, trip log, and vehicle speed. Messages are also sent to/from the driver, and in the event of an unauthorized driver or departure from the authorized region; or upon request from the police, the truck can be remotely disabled.
	Freight Administration (CVO02)	CTI Fleet and Freight Management also coordinates freight transport status, with both intermodal terminals and with the freight equipment (for monitoring the safety and security of the freight).
	Enforcement Case Cooperative Handling (CVO03)	Roadside commercial vehicle inspections are done by the City Management Bureau (CMB) Maintenance and Construction. The roadside inspection stations read the weight of trucks by the axle and also read the carrier ID, vehicle ID, and driver ID. These are compared and analyzed with the historical records for the carrier, vehicle, and driver, as well as existing valid permits and, based on some algorithm, decide which vehicles to pull in for static inspection. When appropriate, the BEDGA Public Transportation Office (PTO) Traffic Enforcement Center is notified for processing violations.
	CV Administrative Processes (CVO04)	CTI Fleet and Freight Management is able to submit credentials applications on behalf of carriers and vehicle owners to the BEDGA PTO Commercial Vehicle Administration Center (CVA). The CVA determines route restrictions for CVs by getting infrastructure restrictions from the CMB Maintenance and Construction Management Center. Route restrictions are posted to the traffic information centers (TICs).

Table A2: continued

ITS Service Area	Service Packages	Functionality
Commercial vehicles (CV) operations	HAZMAT (Hazardous Materials) Management (CVO12)	CTI Fleet and Freight Management keeps track of all HAZMAT shipments. In the event that an equipped commercial vehicle carrying HAZMAT is involved in a crash, the vehicle will send a HAZMAT notification directly to the Emergency Management Office (EMO) Emergency Management Center, including the vehicle and carrier IDs. This information will be sent to the CTI Fleet and Freight Management Center, which will respond with the "bill of lading" that indicates exactly which hazardous materials and how much of them are on the vehicle. This will enable the Fire Department first responders to know exactly what to expect at the crash site, and thus can bring the required equipment for a faster and more effective response.
	Detection of Non-permitted Dangerous/ HAZMAT cargo (CVO13)	Special roadside detection equipment will be deployed by the BEDGA PTO Commercial Vehicle Check Equipment. This may include detectors for nitrogen compounds (indicative of the most common explosives) or for energetic neutrons (indicative of radioactive cargo). The vehicle will also be identified either by its license plate or by its electronic tag. The check stations will then access credentials information from the BEDGA PTO CVA. If a vehicle passes the station sensors and activates them for hazardous materials, and the vehicle does not have the necessary permits, then it will get a message sign to pull in for a static inspection.
	CV Driver Security Authentication (CVO14)	CVs can be preloaded by the CTI Fleet and Freight Management Center with biometric information for each of the authorized drivers of each vehicle. In the event that a driver is not authorized for the vehicle, then the vehicle will not start. Vehicles already started may be remotely disabled.
Data management	ITS Data Warehouse (DM01)	All centers will send real-time operational data to the BEDGA PTO Archived Data System for indexing and storage. The archive administrator will act as a librarian to give users of the archive access to appropriate data. Users will be able to access standard reports, as well as user-defined reports.
Maintenance and construction (M&C)	Maintenance and Construction Vehicle and Equipment Tracking (MC01)	Each CMB Maintenance and Construction Vehicle reports its position to the CMB Maintenance and Construction Management Center dispatch function.

Table A2: continued

ITS Service Area	Service Packages	Functionality
Maintenance and construction (M&C)	Roadway Automated Treatment (MC03)	This service is designed to automatically apply anti-icing chemicals (e.g., liquid magnesium chloride solution) to roadway surfaces that otherwise may be likely to ice based on roadway surface temperature, dew point, and forecasted air temperature.
	Winter Maintenance (MC04)	This service is responsible for surface treatment (anti-icing by maintenance trucks that spread anti-icing chemicals on the roadway) and snow removal by maintenance vehicles configured as snow-plows.
	Roadway Maintenance and Construction (MC05)	CMB and Traffic Management Bureau (TMB) both have ITS Roadway Equipment in the field. These pieces of equipment can report their status to CMB MCMC including faults which will generate "tickets", to CMB Field Maintenance Equipment to repair and clear the faults.
	Work Zone Management (MC06)	The objective of this ITS service is to maintain the safety of ITS work zones where construction or maintenance are being performed.
	Work Zone Safety Monitoring (MC07)	Work zone vehicles in the field have sensors that can passively detect dangerous vehicle behavior (speeding or passing a barrier or crossing a "light curtain") or actively detect vehicles equipped with vehicle-to-infrastructure (V2I) CV technology. Warnings and alarms can be sent immediately to work zone personnel to take evasive action as needed. Work zone staff can be alerted by audible alarms or messages to their personal information devices.
	Maintenance and Construction Activity Coordination (MC08)	CMB MCMC will provide work plans to and collect feedback on those plans from the TMB TMC, BEDGA PTO EMC, EMO EMC, CTI Transit Management Center, and other MCMCs from adjacent regions. The CMB MCMC will also send infrastructure restrictions and asset restrictions to the BEDGA Commercial Vehicle Administration Center, the TIC, Asset Management Systems, and the BEDGA Alternate Mode Transportation Center.
Parking management	Parking Space Management (PM01)	Parking is managed at parking lots in Gui'an by Private Company Parking Management Systems (PMSs). PMSs send parking availability information to the TMB TMC for dissemination on message signs and to the BEDGA PTO TIC for dissemination to travelers on their website, apps, and directly to vehicle's OBE. Connected vehicles can also get parking information using V2I dedicated short-range communications (DSRC) from roadside units.
	Parking Electronic Payment (PM03)	PMSs can receive parking payment from operators using DSRC or wide area wireless communication from the vehicle OBE. The vehicle OBE will access the user payment device for payment information or the payment device can be used directly at the PMS.

Table A2: continued

ITS Service Area	Service Packages	Functionality
Parking management	Regional Parking Management (PM04)	The PMS will share parking availability with the TMC, the CTI Transit Management Center, and the BEDGA PTO TIC. Under this service, the TIC can also make parking reservations for its users/travelers.
Public safety	Emergency Call Taking and Dispatch (PS01)	The EMO Emergency Management Center (EMO EMC) finds out about emergencies from their Public Security Bureau (PSB) Emergency Telecommunications System (emergency call takers at the receiving end of standardized emergency phone number "1-2-0") and the CTI Transit Management Center (for transit-oriented emergencies). The EMO EMC also shares incident reports, and coordinates incident response for significant incidents with other EMCs in adjacent regions.
	Emergency Vehicle Preemption (PS03)	Using the tracking information discussed in CSP PS01 as well as the specific dispatched route for the emergency vehicle to follow, the EMO EMC will send a request for emergency traffic control to the TMB TMC. If granted, then the appropriate signal control commands will be sent to the appropriate TMB ITS Roadway Equipment (the signal controllers) so that the emergency vehicle can pass through the signalized intersections on its route quickly and safely.
	Roadway Service Patrols (PS08)	The purpose of Roadway Service Patrols is to clear disabled vehicles from blocking lanes as quickly as possible. The PSB EMO Emergency Vehicles that serve on this service are dispatched and tracked by the TMB TMC. These service patrol vehicles also provide incident status to the TMC, and the incident records are shared in real time with the EMO EMC, EMB MCMC, and the BEDGA PTO TIC.
	Evacuation and Re-entry Management (PS13)	The objective of this service package is to plan and execute an orderly evacuation and re-entry when necessary, using the transport resources most efficiently. Except with regard to public transportation, the EMO EMC is overall in charge of coordinating resources in an emergency requiring evacuation and subsequent reentry. Public transport resources in Gui'an are coordinated by the BEDGA PTO TMC.
Public transport	Taxi Monitoring and Dispatching (PT01)	The CTI Taxi Management Center tracks the location of CTI taxis, and dispatches them. It also monitors the taximeter data in real time. The Taxi Management Center sends taxi probe data to the TMB TMC and the BEDGA PTO TIC.
	Transit Vehicle Tracking (PT01)	The CTI Transit Management Center tracks the location of CTI transit vehicles and monitors their schedule performance. The Transit Management Center sends transit vehicle probe data to the TMB TMC and BEDGA PTO TIC. The TIC also gets transit vehicle schedule performance.

Table A2: continued

ITS Service Area	Service Packages	Functionality
Public transport	Bicycle Parking Monitoring (PT01)	The Shared Bicycle Operations Center (SBOC) will monitor the location of SBOC-shared bicycles. If bicycles are parked in violation of rules or laws, then the shared bicycle location will be sent to the BEDGA PTO Traffic Operations Monitoring Center to issue a violation.
	Transit Fixed Route Operations (PT02)	The CTI Transit Management Center receives roadway status (congestion and incidents) from the TMC, and infrastructure status from the CMB MCMC. This information may be used to adjust fixed route transit service routes and schedules. Actual routes and schedules (including any adjustments) will be sent to the TIC.
	Public Transportation Fare Collection Management and Industry Supervision (PT04)	Travelers can use a payment instrument or device on board, off-board, or handheld (using an app) to pay for their bus fare. For off-board fare payment, a receipt can be issued (paper or electronic) for verification by on-board inspectors. An app-based payment can have a verification QR code sent to the device, or the verification code or receipt can be read on-board by electronic validators if the purchased fare is a "contract" fare (e.g., "unlimited rides for a month" or a "10-ride ticket").
	Mobile Enforcement Management and Supervision (PT04)	BEDGA PTO Enforcement Management Personnel may carry a BEDGA PTO Mobile Enforcement Device. The personnel will identify and authenticate themselves on the device when logging into the device. The mobile enforcement device can read permits, licenses, and transit passenger fare "contracts." It can also be used for intercity buses and commercial vehicle enforcement with respect to height, weight, and number of allowable passengers. The mobile enforcement device will download from the BEDGA PTO Traffic Enforcement Center dispatch commands and registration information and will respond to the BEDGA PTO Traffic Enforcement Center with dispatch response, electronic evidence information, request for registration information, personnel work attendance information, and time and location information.
	Traffic Enforcement Data Analysis and Evaluation (PT04)	BEDGA PTO Enforcement Management Personnel can send to the BEDGA PTO Traffic Enforcement Center data analysis requests and information review requests. In return, they will get annual assessment results, data analysis results, and information query results.
	Traffic Enforcement Information Bulletin and Service (PT04)	Individuals can use their Personal Information Device (browsers or apps) to do an enforcement query or submit complaints to the BEDGA PTO Traffic Enforcement Center. In response, they will receive case information announcement (or notifications of these messages via a selected channel, e.g., text message, email, app notification) and direct responses to queries and/or complaints.

Table A2: continued

ITS Service Area	Service Packages	Functionality
Public transport	Monitoring and Review of Bus Incidents (PT05)	The CTI Transit Management Center monitors the CTI transit vehicles location, transit vehicle conditions (camera images both internal and external, and audio detected on the bus), and the vehicle's "alarm" buttons (hidden for the driver and public for the passengers).
	Public Transportation Incident Command and Dispatch (PT05)	This service allows buses to be reallocated in response to an emergency elsewhere. Based on the new emergency route, the CTI Transit Management Center can request dynamic bus lanes and traffic control priority from the TMB Traffic Management Center
	Transit Security (PT05)	The BEDGA PTO EMC and the EMO Emergency Management Center can send incident information and threat information to the CTI Management Center.
	Transport Hub Operation Monitoring and Management (PT05)	The Transport Hub Management Center sends and receives transport demand information, incident information, passenger flow information, emergency data, and decision information.
	Taxi Industry Supervision (PT06)	The CTI Taxi Management Center can issue violation behavior warnings to the CTI Taxi OBE, which is displayed to the operator. The taxi OBE sends vehicle emissions, location data, and taxi meter data to the CTI Taxi Management Center.
	Transit Passenger Counting (PT07)	The CTI Transit Vehicle OBE uses technology to count the passengers entering/exiting the bus, and sends the transit vehicle loading data to the CTI Transit Management Center. Ridership is then reported to personnel at both CTI Transit Operations and BEDGA PTO Transit Management.
	Transit Traveler Information (PT08)	Travelers can access transit (and other travel information) from the TIC using either their Personal Information Device or the Traveler Support Equipment located at Hubs and other popular transit locations.
	Transit Signal Priority (PT09)	CTI Transit Vehicles OBE reports their position and schedule performance (i.e., by how much they are behind schedule) to the CTI Transit Management Center. At the same time, the TMB TMC will have closed-loop signal control (i.e., they can remotely change the traffic signal timing plans at any intersection).
	Bus Lane Management (PT10)	This service allows a lane to be dedicated exclusively to bus traffic, either during regularly scheduled times or when buses are running behind schedule due to congestion.

Table A2: continued

ITS Service Area	Service Packages	Functionality
Public transport	Multimodal Coordination (PT14)	The CTI Transit Management Center coordinates with other transit management centers by sharing schedules and making schedule adjustments so that passengers can more easily make connections between modes. The CTI Transit Management Center also sends their transit system data to the TMB TMC, gets parking information from the Private Company Parking Management Systems, exchanges service information with the BEDGA PTO Alternate Mode Transportation Center, and makes this information available to travelers on CTI transit vehicles by sending it to buses for message sign display and annunciation.
	Supervision of the Bicycle Sharing Industry (PT14)	The BEDGA PTO Transit Management Personnel can request and receive operation reports from the SBOC-shared Bicycle Operation Center, and then provide review comments to the SBOC-shared Bicycle Operation Center.
	Integrated Multimodal Electronic Payment (PT18)	The CTI Payment Administration Center receives payment information from Vehicle OBE and Personal Information Devices for transit and transit-related services. The payments are forwarded to Private Company Parking Management Systems (for parking) and Traveler Support Equipment (for transit ride "contracts"). The payments are reconciled by the CTI Financial Center and payments may be coordinated with other payment administration centers.
Sustainable travel	Roadside Lighting (ST04)	The TMB TMC detects vehicles using V2I-connected vehicle technology and, based on vehicle location and direction, will turn on (or set to full illumination) roadway lighting as needed. When vehicles are not present, roadway illumination can be turned off or set to a lower level of illumination.
	Electric Charging Stations Management (ST05)	Electric vehicles either looking for charging or scheduling charging can provide their vehicle charging profile (how long they will need to charge and the charge power that their vehicle can accommodate) to the TIC, and receive (available) electric charging services inventory from the TIC. This exchange with the TIC can be direct over wide area wireless (e.g., GSM) or via V2I BEDGA PTO Connected Vehicle Roadside Equipment.

Table A2: continued

ITS Service Area	Service Packages	Functionality
Traveler information	Personalized Traveler Information (TI02)	The TIC offers traveler information over a Traveler Information Voice Response System and provides traveler information to the media. The TIC also provides interactive traveler information through a smartphone app and at kiosks (Traveler Support Equipment). If using an app for traveler information, travelers may get alerts if they choose and if the conditions of their trip have changed while they are traveling.
	Infrastructure Provided Trip Planning & Route Guidance (TI04)	When a traveler selects a route provided by the TIC, the TIC may indicate the "logged vehicle route" to the TMC and/or "trip confirmation" to the CTI Transit Management Center. While this may or may not constitute a formal reservation, it may allow the roadway operator and/or the transit service provider to plan for specific demand in the future. In this mode, the TIC may provide the traveler several alternative trips, and the traveler will select one. "Reservation-based" traveler information can be done by a traveler from smartphone app, a kiosk, or vehicle OBE.
	Information Service at Hub Station	At hub stations, the TIC will support travelers to make parking reservations at the Private Company Parking Management Systems, make electronic payments with the Finance Bureau Financial Center, support alternate mode information from the BEDGA PTO Alternative Mode Transportation Center, and make general travel services reservations from the BEDGA PTO Travel Services Provider System. Travelers will be able to access hub travel services at the TIC using their smartphone app or kiosks.
Traffic management	Infrastructure-Based Traffic Surveillance (TM01)	This CSP uses video- and detector-based surveillance of roadways to analyze congestion and support detection/classification/response to incidents by the Traffic Management Bureau's Traffic Management Center (TMB TMC or TMC for short). The TMC shares road network conditions (e.g., link congestion and nonrecurring incidents) with the BEDGA PTO TIC (Transportation Information Center).
	Connected Vehicle-Based Traffic Surveillance (TM02)	The TMC can augment vehicle-based surveillance with buses (from the CTI Transit Management Center), toll roads (from CTI Payment Administration Center), and the TIC (which may implement a "WAZE" like smartphone app-based vehicle guidance/tracking service).
	Traffic Signal Control (TM03)	Implements traditional actuated signal control with TMC-based coordination and oversight. In addition to vehicle detection, detect bicycles and pedestrians to include their intersection demands in the signal optimization process.

Table A2: continued

ITS Service Area	Service Packages	Functionality
Traffic management	Connected Vehicle Traffic Signal System (TM04)	Includes traffic situational data derived from BEDGA PTO-connected vehicle roadside units as input to the traffic signal control.
	Traffic Metering (TM05)	Signals will be placed at the end of ramps feeding limited access highways, and these signals will be used to "meter" the on-ramp vehicles onto the mainline, so that the mainline vehicle density remains on the left of the "flow-density" curve (i.e., to avoid stop-and-go congestion of the limited access highway). During times of low traffic density, these meters will be disabled or "green" all the time.
	Traffic Information Dissemination (TM06)	The TMC will disseminate traffic information by message signs along the roadways, and to other centers, e.g., EMO Emergency Management Center, CMB Maintenance and Construction Management Center, CTI Transit Management Center, and BEDGA PTO TIC. The TIC will produce "traveler information for media" based on the needs of television and radio broadcast stations for their "traffic reports." .
	Traffic Incident Management System (TM08)	The EMO Emergency Management Center (EMC) will share incident information it logs with other centers, as well as receive incident records from these centers (traffic, transit, TIC, maintenance and construction, PSB Border Inspection System, and the Rail Operations Center). The EMO can deploy appropriate vehicles from its emergency vehicle fleet, and get additional incident information from the emergency vehicle operators after they have arrived at the incident scene. The TMC will operate the video surveillance system (e.g., high-resolution cameras with PTZ control) to verify and validate the severity of incidents on the roadways under such surveillance. This information will be shared with the TIC and PSB Border Inspection Systems as appropriate.
	Integrated Decision Support and Demand Management. (TM09)	The TMC will be able to influence, control, or cooperate with (depending on interagency policy agreements) operational policies with other centers.

Table A2: continued

ITS Service Area	Service Packages	Functionality
Traffic management	Traffic Comprehensive Monitoring (TM09)	The BEDGA PTO Traffic Operations Monitoring Center will collect and aggregate incident information from the EMC, roadway maintenance status from the Maintenance and Construction Management Center, transit system data from the CTI Transit Management Center, parking information from the Private Company Parking Management Systems, logged vehicle routes from the TIC, air quality information from the BEDGA PTO Traffic Emissions Management Center, transit system data from the CTI Taxi Management Center, and event plans from the BEDGA PTO Event Promoter System.
	Dynamic Roadway Warning (TM12)	Field equipment with a combination of cameras and detectors, under TMB Traffic Management Center remote operator control, will warn drivers of hazards or operational changes by message signs.
	Speed Warning and Enforcement (TM17)	Field equipment with vehicle speed detectors, under TMB Traffic Management Center remote control, will warn drivers of their speed as well as send speed monitoring and speed violation information to the BEDGA PTO Traffic Enforcement Center. The speed warning to drivers may be by message sign or via a roadside unit to in vehicle annunciation or display function. The BEDGA PTO Traffic Enforcement Center may also control the speed violation notification function (e.g., set speed thresholds for notification).
Vehicle safety	Road Weather Motorist Alert and Warning (VS07)	The TMC will collect and analyze information from the Weather Service System as well as the BEDGA PTO Surface Transportation Weather Service, the CMB ITS Roadway Equipment, and the TIC based on its interface with individual vehicles that provide vehicle environmental data; and disseminate road network conditions (including weather information) to the TIC, roadside units (to send to passing vehicles), Transit Management Center, and the CMB Maintenance and Construction Management Center.

Source: https://www.itsknowledgeresources.its.dot.gov/its/bcllupdate/pdf/BCLL_TRANSIT_INFO_2017_FINAL.pdf.

Appendix 3

Table A3: Potential Benefits of Selected Gui'an Customized Service Packages

Potential Benefits / Gui'an Customized Service Packages	TI05: Information Service at Hub Station	TM01: Infrastructure-Based Traffic Surveillance	TM02: Vehicle-Based Traffic Surveillance	TM03: Traffic Signal Control	TM04: Connected Vehicle Traffic Signal System	TM05: Traffic Metering
Transit traveler information increased new or more frequent "choice" riders 40%–70% decreased paratransit no-shows 45%, increased en-route mode shift to transit 2%–8%, increased perception of service reliability 64% (https://www.itsknowledgeresources.its.dot.gov/its/bcllupdate/pdf/BCLL TRANSIT_INFO_2017_F	X					
Traffic Incident Management Systems reduced: incident duration 12%–58%, secondary crashes 69%, incident validation time 50%–80%, roadway clearance time 11%-18%, incident clearance time 4%. Service Patrols increased customer satisfaction 93%–95% (https://www.itsknowledgeresources.its.dot.gov/its/bcllupdate/pdf/BCLL_TrafficIncidentManagement_2017_FINAL.pdf).						
Work Zone automated speed enforcement reduced speeding vehicles by 50%, Work Zone Information systems reduced delay by 46%–55%						
Integrated Corridor Management (2011-2012) shows improvements in travel time reliability of 3.0%-10.6%		X	X	X		X
A ramp meter evaluation in Kansas City (2012-13) show reduction in crashes 26%–50%, reduce crashes by 64%, increase corridor throughput by 20%, reduce incident clearance by avg. 4-minutes, and reduce peak travel times by 1%–4% (https://www.itsknowledgeresources.its.dot.gov/its/bcllupdate/pdf/BCLL_Freeway_Overview_2017_FINAL.pdf).						X
8.2% NHTSA estimates that safety applications enabled by V2V and V2I could eliminate or mitigate the severity of up to 80% (depending on CV market penetration) of non-impaired crashes, including crashes at intersections or while changing lanes (https://www.itsknowledgeresources.its.dot.gov/its/bcllupdate/pdf/BCLL%20CV%20Safety%202018%20v2.pdf).				X		
Transit Signal Priority operations (Anthem, Arizona, 2015) shown to improve connected bus travel times by 8.2%				X		
Modeling results when AVs with CV capabilities implement V-I and are programed to minimize environmental costs: fuel consumption - (5-10)%-13% on a coordinated corridor.			X	X	X	
AV reductions: travel time 12%–48%, delay 0%–85%, emissions 40%–50%						
Adaptive traffic signal systems: travel time reduction 1%–53%, increase in avg. speed 8%–25%, fuel consumption reduction 3%–8%, emissions reduction 0%–17%, stop reduction 22%–34%, delay reduction 10%–40%, crash reduction 29% (https://www.itsknowledgeresources.its.dot.gov/its/bcllupdate/pdf/BCLL_TrafficControl_2017_FINA L.pdf).		X		X	X	
Arterial signal control: travel time reduction 8%–35%, increase avg. speed 8%–18%, reduce fuel consumption 5%–14%, emissions reduction 8%–15%, stop reduction 11 75%, delay reduction 6%–93%, crash reduction 28%–31% tsknowledgeresources.its dot.gov/its/bcllupdate/pdf/BCLL_TrafficControl_2017_FINAL.pdf).		X		X	X	
In Salt Lake City, Utah, cumulative travel time reduced 4.3% by deploying the weather responsive timing plans			X			
In St. Paul, Minnesota, an advanced parking management system reduced travel times by 9%						
Speed warning and enforcement showed reductions in: crash injuries 7%–98%; crashes 5%–65%; fatalities 8%–82%; speed 10%; speeding vehicles 58%						

Potential Benefits / Gui'an Customized Service Packages	TM06: Traffic Information Dissemination	TM08: Traffic Incident Management System	TM09: Integrated Decision Support and Demand Management	TM17: Speed Warning and Enforcement	WX01: Weather Data Collection	WX02: Weather Information Processing and Distribution	WX03: Spot Weather Impact Warning
Transit traveler information increased new or more frequent "choice" riders 40%–70% decreased paratransit no-shows 45%, increased en-route mode shift to transit 2%–8%, increased perception of service reliability 64% (https://www.itsknowledgeresources.its.dot.gov/its/bcllupdate/pdf/BCLL TRANSIT_INFO_2017_F		X					
Traffic Incident Management Systems reduced: incident duration 12%–58%, secondary crashes 69%, incident validation time 50%–80%, roadway clearance time 11%–18%, incident clearance time 4%. Service Patrols increased customer satisfaction 93%–95% (https://www.itsknowledgeresources.its.dot.gov/its/bcllupdate/pdf/BCLL_TrafficIncidentManagement_2017_FINAL.pdf).	X	X					
Work Zone automated speed enforcement reduced speeding vehicles by 50%, Work Zone Information systems reduced delay by 46%–55%	X	X		X			
Integrated Corridor Management (2011-2012) shows improvements in travel time reliability of 3.0%-10.6%	X	X	X				
A ramp meter evaluation in Kansas City (2012-13) show reduction in crashes 26%–50%, reduce crashes by 64%, increase corridor throughput by 20%, reduce incident clearance by avg. 4-minutes, and reduce peak travel times by 1%–4% (https://www.itsknowledgeresources.its.dot.gov/its/bcllupdate/pdf/BCLL_Freeway_Overview_2017_FINAL.pdf).							
8.2% NHTSA estimates that safety applications enabled by V2V and V2I could eliminate or mitigate the severity of up to 80% (depending on CV market penetration) of non-impaired crashes, including crashes at intersections or while changing lanes (https://www.itsknowledgeresources.its.dot.gov/its/bcllupdate/pdf/BCLL%20CV%20Safety%202018%20v2.pdf).							
Transit Signal Priority operations (Anthem, Arizona, 2015) shown to improve connected bus travel times by 8.2%							
Modeling results when AVs with CV capabilities implement V-I and are programed to minimize environmental costs: fuel consumption -(5-10)%-13% on a coordinated corridor.							
AV reductions: travel time 12%–48%, delay 0%–85%, emissions 40%–50%							
Adaptive traffic signal systems: travel time reduction 1%–53%, increase in avg. speed 8%–25%, fuel consumption reduction 3%–8%, emissions reduction 0%–17%, stop reduction 22%–34%, delay reduction 10%–40%, crash reduction 29% (https://www.itsknowledgeresources.its.dot.gov/its/bcllupdate/pdf/BCLL_TrafficControl_2017_FINA L.pdf).							
Arterial signal control: travel time reduction 8%–35%, increase avg. speed 8%–18%, reduce fuel consumption 5%–14%, emissions reduction 8%–15%, stop reduction 11 75%, delay reduction 6%–93%, crash reduction 28%–31% tsknowledgeresources.its dot.gov/its/bcllupdate/pdf/BCLL_TrafficControl_2017_FINAL.pdf).							
In Salt Lake City, Utah, cumulative travel time reduced 4.3% by deploying the weather responsive timing plans					X	X	X
In St. Paul, Minnesota, an advanced parking management system reduced travel times by 9%							
Speed warning and enforcement showed reductions in: crash injuries 7%–98%; crashes 5%–65%; fatalities 8%–82%; speed 10%; speeding vehicles 58%				X			

Table A3: continued

AV = automated vehicle, CV = commercial vehicle, NHTSA = National Highway Traffic Safety Administration, V2I = vehicle-to-infrastructure, V2V = vehicle-to-vehicle.(Note: Please fill in missing spelled out form.)
Source: Asian Development Bank. 2019. *Gui'an New District ITS Conceptual Design.* Consultant's Draft Final Report. Manila (TA9437-PRC).

Appendix 4

Project Communications Requirements Estimates

The project design team determined by making high-level estimates and assumptions that the communications channels between the stakeholder ITS elements in the CSPs can be deployed in the Gui'an region with a combination of three communication technologies. The three communication technologies selected are dedicated short-range communications (DSRC), Long-Term Evolution (LTE) wireless data (as currently or soon to be deployed in the region and the PRC, generally), and BEDGA's own dedicated fiber-optic network.

In all cases, the speeds and latencies required to make the associated ITS services adequately functional are relatively modest compared with the speeds and latencies currently available in the three communication technologies. DSRC is the wireless data radio technology developed specifically for vehicle-to-vehicle, vehicle-to-NMT (nonmotorized transport), vehicle-to-infrastructure, and NMT-to-infrastructure communications. The key services here are for safety (e.g., collision avoidance), and the communication requirements are very low latency, high speed, and fast establishment of a connection between and with moving platforms. Security is also a major consideration, and the necessary security methods for authentication of mobile and fixed "stations" (using authentication certificates) have been standardized and tested. Today, this is the only commercially available technology that securely meets the technical and security requirements for the safety-critical services.

LTE wireless is the data service that supports wireless mobile services for today's (and the foreseeable future) smartphones. It is a wireless extension of intnernet connectivity and is more than adequate to support many ITS services (but NOT the ITS safety services). This technology is mature, and there is little risk that this service will be deployed and will serve many ITS services. LTE wireless will be able to seamlessly interoperate with conventional 5G data services when they are initially deployed.

Dedicated fiber-optic network, deployed by BEDGA, will be used to connect field equipment to the data center (and will be the highest speed channel to support services such as high-resolution video surveillance). It will also be used to connect transportation management centers that are not located in the BEDGA data center. This technology is mature, and there is little risk that this service will be deployed and will serve many ITS services.

Table 9 summarizes the results of this analysis. In the DSRC column, the services that make use of DSRC also indicate if the modality is vehicle-to-infrastructure (V2I/I2V) or vehicle-to-vehicle (V2V). Note that "V" could also be replaced by "NMT" (nonmotorized travel, i.e. bicycle or pedestrian). In the LTE wireless column, an entry indicates that LTE wireless data is required, and the necessary latency (longest delay time for a connection to be established or for the data sent to be delayed before being received) is indicated. Again, the latency requirements (as well as the data speed requirements) are all relatively modest for normal LTE wireless service capabilities. Finally, the BEDGA fiber-optic network shows the expected data rates required, and these range from a low of 200 megabytes per second (MB/sec) to 1000 MB/sec, all relatively modest speeds for a modern fiber-optic network with conventional routers.

Table A4: Communications Requirement for Selected Gui'an Customized Service Packages

Telecom Channels Customized Service Packages	DSRC	LTE Wireless (latency less than X sec)	BEDGA Fiber Optic (speed in MB/sec)
CVO01: Truck Operation Monitoring		30 Sec	
CVO02: Freight Administration		30 Sec	500
CVO03: Enforcement Case Cooperative Handling		15 Sec	500
CVO12: HAZMAT Management		15 Sec	500
CVO13: Detection of Non-Permitted Dangerous/ HAZMAT Cargo	V2I/I2V	15 Sec	300
CVO14: CV Driver Security Authentication	V2I/I2V	15 Sec	
CVO15: Fleet and Freight Security		15 Sec	
DM01: ITS Data Warehouse			1,000
TM01: Infrastructure-Based Traffic Surveillance		15 Sec	
TM02: Vehicle-Based Traffic Surveillance	V2I/I2V	15 Sec	
TM03: Traffic Signal Control			500
TM04: Connected Vehicle Traffic Signal System	V2I/I2V	15 Sec	500
TM05: Traffic Metering		15 Sec	
TM06: Traffic Information Dissemination			800
TM08: Traffic Incident Management System			500
TM09: Integrated Decision Support and Demand Management			1,000
TM09: Traffic Comprehensive Monitoring			1,000
TM12: Dynamic Roadway Warning			500
TM17: Speed Warning and Enforcement	V2I/I2V		500
VS07: Road Weather Motorist Alert and Warning	V2I/I2V		300
WX01: Weather Data Collection	V2I/I2V	15 Sec	300
WX02: Weather Information Processing and Distribution			300
WX03: Spot Weather Impact Warning	V2I/I2V		500

BEDGA = Bureau of Economic Development in Gui'an New District, CV = commercial vehicle, DSRC = dedicated short-range communications, HAZMAT = Hazardous Materials, LTE = Long-Term Evolution, MB/sec = megabytes per second, V2I = vehicle-to-infrastructure, V2V = vehicle-to-vehicle.

Source: Asian Development Bank. 2019. *Gui'an New District ITS Conceptual Design*. Consultant's Draft Final Report. Manila (TA9437-PRC).

Appendix 5

Open Standards and Protocols Recommendations

Appendix A5 shows for each CSP the relevant application layer open standards and/or protocols that should be considered when designing the ITS elements for that CSP. These standards define the data elements in a data dictionary, and define messages between ITS subsystems as collections of data elements. The use of open standards for the application layer will reduce the risk of interface incompatibility, especially where the elements sharing information are deployed at different times and by different stakeholders. For example, the Traffic Management Bureau Traffic Management Center (TMB TMC) operated by the Traffic Police will not be built as part of the ADB-funded BEDGA ITS project. However, there are many important information interfaces to the future TMB TMC to the other project elements included in the ADB-funded BEDGA ITS project.

For these information flows to be implemented later after the TMB TMC and BEDGA ITS projects are completed, the relevant information flows must be encoded using identical design specifications. By associating the information flows in the CSPs that include both the TMB TMC and information flows to/from elements of the BEDGA ITS project using the same open standard, the risk of interoperability failure when the systems are interconnected is reduced substantially.

Note that there are other standards supporting the transport of the messages for the application layer standards. As long as internet-based standards are used for the message transport, there should be no problem in using a variety of transports, because these transport standards can be translated in conventional routers between the different internet standards. In particular, the transport standards in common usage for mobile wireless service (e.g., GSM 4G LTE) or data over fiber optic or Ethernet services should work fine for transporting and routing these application-specific standardized messages between ITS elements of the same or different stakeholders.

The one exception here is vehicle-to-X where "X" is either another vehicle or a roadside element or a traveler using NMT (bicycle or pedestrian). In this case, it is essential that all elements use the same message transport as well as the same data dictionary and message sets. This is because the vehicles may be communicating directly with another vehicle or pedestrian or bicycle, and there would be no intervening router to make the protocol translation. Thus vehicle-to-vehicle communication must use the same standards/protocols for the entire communication stack.

Of course, it must be noted that simply specifying an ITS open standard is not sufficient to assure interoperability once the elements sharing information are connected. The ITS element sending a message to another ITS element must each implement the open standard governing the interface using exactly the same "optional" objects from the open standard. The specification of which optional objects from the selected open standard to specify for an individual interface is part of the high-level and/or detailed design process that will be conducted by the selected design institute for each of the ITS projects. Each design institute must document and include in their Interface Control Document their selected optional objects that will be implemented, and these selections must be identical between the two elements that will share information. The open standards generally specify a few mandatory objects that each element supporting a standardized interface must implement, and also list optional objects that can be selected for implementation. Each optional object adds a functional capability. The design institute analysts can study the list of optional objects, and select only those functional capabilities that are necessary for the proper functioning of the projects.

Selecting Between Alternative Standards or Protocols

In one case in Appendix A5, there are alternative open standards/protocols that each encode very close to the same information, but differently: CEN-SIRI Transit Schedule versus GTFS-Real-time Transit Schedule.Both standards (technically, GTFS-RT is a protocol because it is not governed by an open standards committee) can encode bus schedules and the reported schedule can include real-time deviations from the schedule for a bus run that is in progress. They both allow for services that use the real-time schedule deviation of the bus, for example, to provide information to a traveler about exactly when a bus is expected to arrive at a bus stop (the travelers start or end of a bus ride). In Europe, the CEN-SIRI standard is used predominantly by public transit agencies. In North America, both standards are used, with many more deployments of GTFS-RT (but with some notable exceptions using CEN-SIRI, for example, both MTA Transit in the New York City region and Metropolitan Transportation Commission in the San Francisco Bay Area). In Gui'an, the decision might be made by which standard is used predominantly in the PRC. This decision would likely give more equipment options for the selected design institute.

For communications between a vehicle and an infrastructure center, it is assumed that the vehicle will use commercially available GSM 4G LTE data communications, which can be viewed as internetworking with the internet. For communications with other vehicles/bicycles/pedestrians/road-side infrastructure for safety applications, the message delay or latency of GSM 4G LTE makes use of this service unacceptable. Not shown in Table 10, because these are not application layer standards, is the choice for vehicle-to-X communications (where "X" can be for example V= another vehicle, i= infrastructure, p= pedestrian, b= bicycle). Today, there is only one operational standard ready for deployment: the IEEE 1609 - family of standards for Wireless Access in Vehicular Environments (WAVE). These standards have been available since 2013, chipsets to support their implementation are available today, and many of the safety applications have been tested and found to be effective. General Motors is already deploying these standards in one version of their "Cadillac" automobile, and other vehicle manufacturers have committed to deploying this standard.

Based on the stage of development for C-V2X, the project team's recommendation is to use the IEEE 1609 DSRC standard for any V2X applications in the next few years.

Table A5: Application Layer Open Standards and Protocols Associated with Relevant Customized Service Packages

Customized Service Packages:	ANSI TS813 CVO E-Filing	APTA TCIP Transit Internal	ASTM E2468-05-Metadata Archive	ASTM E2665-08 Archive Data	CEN-SIRI Trnst Sched+	GTFS-Realtime Trnst Sched	IEEE 1512-2006 Incident Management	IEEE 1609.11 E-Payment	ITE TMDD – Traffic Mgmt Data Dict	NTCIP 1201-GO	NTCIP 1202-ASC	NTCIP 1203-DMS	NTCIP 1204-ESS	NTCIP 1205-CCTV	NTCIP 1206-DCM	NTCIP 1207-RM	NTCIP 1208-VS	NTCIP 1209-TSS	NTCIP 1210-FMS	NTCIP 1211-SCP	NTCIP 1213-ELMS	SAE J2354-ATIS	SAE J2735 DSRC	SAE J2945/1 V2V Safety	SAE J3067 J2735 Update
CVO01: Truck Operation Monitoring									X																X
CVO02: Freight Administration																									X
CVO03: Enforcement Case Cooperative Handling																									X
CVO12: HAZMAT Management							X																		X
CVO13: Detection of Non-Permitted Dangerous/ HAZMAT Cargo																									X
CVO14: CV Driver Security Authentication																									X
CVO15: Fleet and Freight Security																									X
DM01: ITS Data Warehouse			X	X		X	X		X			X		X			X					X	X		
TM01: Infrastructure-Based Traffic Surveillance									X		X			X				X	X	X					
TM02: Vehicle-Based Traffic Surveillance				X					X														X	X	
TM03: Traffic Signal Control										X	X			X				X	X	X					
TM04: Connected Vehicle Traffic Signal System										X	X							X	X	X			X	X	
TM05: Traffic Metering										X	X			X		X	X	X	X						
TM06: Traffic Information Dissemination									X		X	X						X							
TM08: Traffic Incident Management System							X		X		X			X				X	X	X					
TM09: Integrated Decision Support and Demand Management	X								X		X											X			
TM09: Traffic Comprehensive Monitoring	X								X		X											X			
TM12: Dynamic Roadway Warning										X	X			X				X	X	X					
TM17: Speed Warning and Enforcement											X						X						X		
VS07: Road Weather Motorist Alert and Warning									X				X										X		
WX01: Weather Data Collection									X				X										X		X
WX02: Weather Information Processing and Distribution									X																
WX03: Spot Weather Impact Warning									X			X	X				X						X		

CV = commercial vehicle, HAZMAT = Hazardous Materials, ITS = intelligent transport systems.
Source: Asian Development Bank. 2019. *Gui'an New District ITS Conceptual Design.* Consultant's Draft Final Report. Manila (TA9437-PRC).

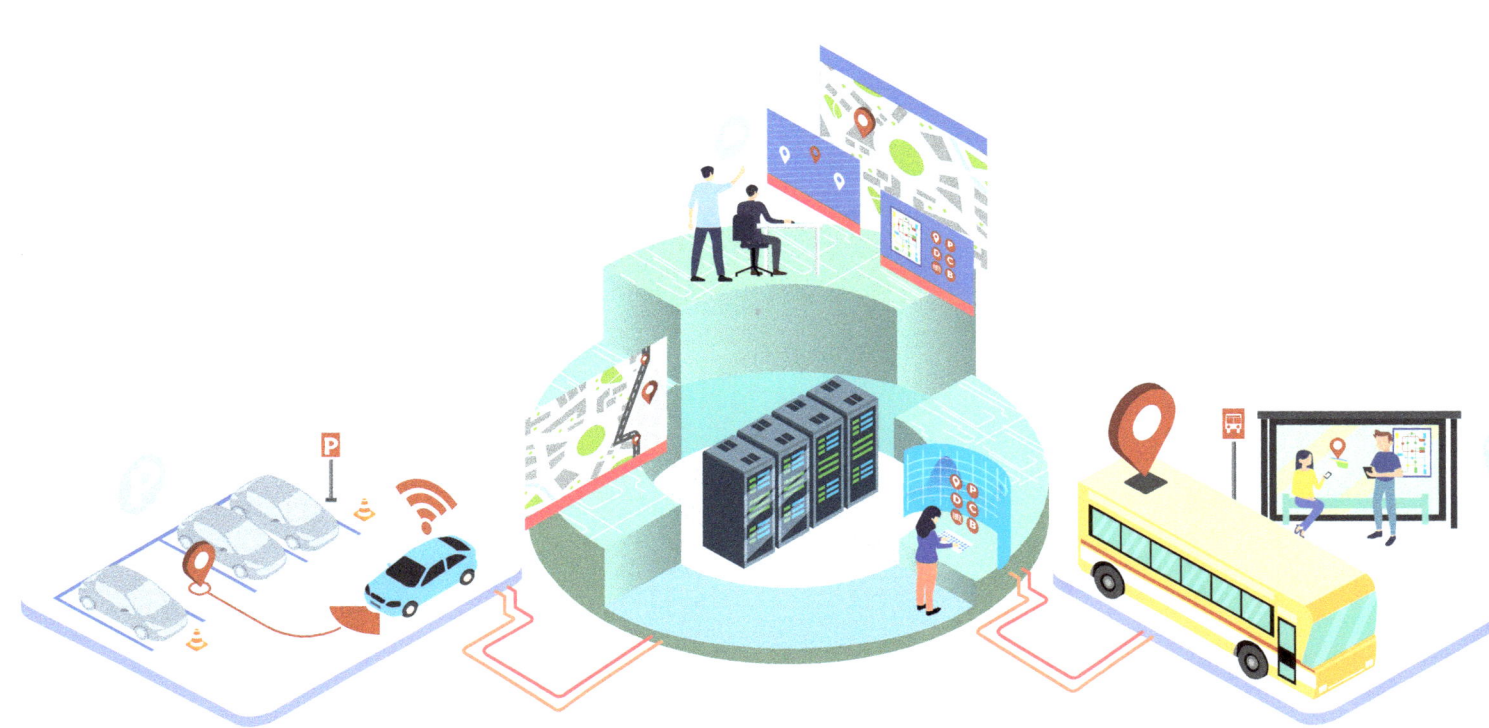

Lightning Source UK Ltd.
Milton Keynes UK
UKHW050744031020
370924UK00007B/142